D0436973

THE PHYSICS OF CHRISTMAS

Also by Dr. Roger Highfield

FRONTIERS OF COMPLEXITY,
with Peter Coveney

THE PRIVATE LIVES OF
ALBERT EINSTEIN,
with Paul Carter

THE ARROW OF TIME,
with Peter Coveney

The PHYSICS of CHRISTMAS

From the Aerodynamics of Reindeer to the Thermodynamics of Turkey

Dr. Roger Highfield

LITTLE, BROWN AND COMPANY
Boston New York London

Copyright © 1998 by Roger Highfield

All rights reserved. No part of this book may be reproduced in any form or by any electronic or mechanical means, including information storage and retrieval systems, without permission in writing from the publisher, except by a reviewer who may quote brief passages in a review.

FIRST EDITION

The author is grateful for permission to include the following previously copyrighted material: "Talking Turkeys," copyright © Benjamin Zephaniah, 1994. Taken from the book *Talking Turkeys,* published by Penguin Books.

LIBRARY OF CONGRESS CATALOGING-IN-PUBLICATION DATA
Highfield, Roger.
The physics of Christmas : from the aerodynamics of reindeer to the thermodynamics of turkey / Roger Highfield. — 1st ed.
p. cm.
Includes bibliographical references and index.
ISBN 0-316-36611-0 (hardcover).
1. Christmas. 2. Science — Miscellanea. I. Title.
GT4985.H53 1998
394.2663 — dc21 98-20072

10 9 8 7 6 5 4 3 2

MV-NY

Book design Steve Dyer

Printed in the United States of America

CONTENTS

IN MEMORY OF MY FATHER

This will be my first Christmas without you.

ACKNOWLEDGMENTS

I have endeavoured in this Ghostly little book, to raise the Ghost of an Idea, which shall not put my readers out of humour with themselves, with each other, with the season, or with me. May it haunt their house pleasantly, and no one wish to lay it.

Their faithful friend and servant,
C.D.
December 1843

Like Charles Dickens in his preface to *A Christmas Carol,* I propose "to raise the Ghost of an Idea."

In Dickens's classic book, the Ghosts of Christmas Past, Present, and Future reveal the true meaning and spirit of the season to Ebenezer Scrooge, transforming him from a miser into a potent symbol of charity. I, too, hope to enlighten the reader by acting as a guiding spirit, one who will illuminate Christmas by viewing the holiday and its rituals from a new perspective, that of science. Christmas and associated celebrations offer a wonderful excuse to explore a broad range of fields, from biotechnology and fractals to neuropharmacology and nanotechnology. If appetites are whetted for science, or at the very least curiosity about the subject is stimulated, I will be pleased. Any change in the charitable behavior of the reader would, of course, be a welcome bonus.

Each Christmas for the past decade, I have written about seasonal science for *The Daily Telegraph.* Many thanks to my editor, Charles Moore, and his predecessor, Max Hastings, for indulging my obsession. Gulshan Chunara, as ever, provided me with invaluable assistance. It has also been stimulating discussing aspects of the book with my colleagues Aisling Irwin, David Johnson, Laura Spinney, Tom Standage, and Robert Uhlig. Sarah Foot provided me with a wonderful retreat on Islay where I could work on the U. S. edition.

John Brockman and Katinka Matson gave me the encouragement to develop a book proposal. Very many thanks are also due to Little, Brown and Rowohlt for backing the project and in particular to Rick Kot for his warm encouragement and support. It has also been a pleasure to work with Susanne McDadd of Metro, London, on a third book project together.

I would like to thank the following people and organizations for helping me to seek out the science behind the festivities: Leonard Adleman, Denis Alexander, Anthony Astbury, Peter Barham, Linda Bartoshuk, Sam Berry, Gerard Bond, Roy Bradshaw, British Antarctic Survey, Donald Brownlee, L. P. Bucklin, Stephen Burley, David Cheal, David Clary, Chris Clayton, Cary Cooper, Malcolm Cooper, Winnifred Cutler, Daniel Dietrich, Piero Dolara, Jonathan Dorfman, Robert East, Matthew Freeman, Jeffrey Friedman, Adrian Furnham, Takanari Gotoda, Richard Gross, Steven Guest, Sunil Gupta, Odd Halvorsen, Patrick Harding, Alan Hirsch, James Horne, David Hughes, Ilpo Huhtaniemi, Colin Humphreys, Nina James, Steve Jones, Dan

Keathley, David Kelly, Barry Kemp, Harold Koenig, Sir Harry Kroto, Michel Laroche, Dale Lewison, Robin Lovell-Badge, Laurie Lucchina, George Masterton, Dale Matthews, John Maynard Smith, Patrick McGovern, Wendy Mechaber, Dave Mela, John Metz, Michael Molnar, John Moore, Les Noble, Adrian North, Jose Pardo, Kenneth Pargament, David Peel, Raj Persaud, David Phillips, Krishna Podila, Bill Proebsting, Sir Martin Rees, Wolf Reik, Allen Riordan, Norman Rosenthal, Delwen Samuel, Nigel Scott, Larry Silverberg, David Skuse, Kristina Staley, Ian Stewart, Andrew Strassman, Joergen Taageholt, Fred Turek, Luca Turin, Mark Uncles, UNICEF, Alan Watkins, Diederik Wiersma, George Williams, and Ian Wilmut.

Ronald Parkinson of the Victoria and Albert Museum in London was kind enough to spend a morning with me discussing the museum's vast collection of Christmas cards. Linda Capper also proved a great help when it came to contacting members of the British Antarctic Survey.

A number of people have also read parts or all of the manuscript to ensure that the science is understandable. Many thanks to my wonderful wife, Julia; my parents, Ron and Doris; and a number of friends: Samira Ahmed, Peter Coveney, Tony Manzi, Eamonn Matthews, Brian Millar, Sharon Richmond, and Martin Winn. I'd like to thank the Reverend Dr. John Platt of Pembroke College, Oxford, for looking over the chapter on the Star of Bethlehem.

Particular thanks to Graham Farmelo of the Science Museum for his many and constructive suggestions on an early draft, to Robert Matthews for his calculating skills and his Murphy's Law expertise, and to my mother for translating German-language papers and Christmas books.

I am also indebted to a number of researchers who gave

me feedback on specific sections of the book. I have covered such a wide range of fields that I am confident of one thing: a number of howlers remain, all of which are my responsibility. Many thanks to the following for helping me weed out some of the worst: Miguel Alcubierre, Peter Atkins, Peter Barham, Sam Berry, David Bonthron, Roy Bradshaw, Roger Buckland, Carole Burgoyne, Linda Capper, David Clary, Roger Cone, Cary Cooper, Peter Coveney, Glenn Cox, Peter Davies, Daniel Dietrich, Robert East, Sabine Eber, Ron Evans, Matthew Freeman, Adrian Furnham, David Gems, Alexei Glebov, Richard Gross, Rose Gubitosi-Klug, Sunil Gupta, Laurance Hall, Odd Halvorsen, Patrick Harding, James Horne, David Hughes, Ilpo Huhtaniemi, Colin Humphreys, Dan Keathley, David Kelly, Gerd Kempermann, Harold Koenig, Tom Lachlan-Cope, Michel Laroche, Dale Lewison, Robin Lovell-Badge, Neil Martin, Patrick McElduff, Stanley McKnight, Dave Mela, Randolf Menzel, Daniel Miller, Les Noble, Adrian North, José Pardo, Daniele Piomelli, Caroline Pond, David Price-Williams, Wolf Reik, Allen Riordan, Margaret Robins, Delwen Samuel, Larry Silverberg, Gene Stanley, Ian Stewart, Scott Swartzwelder, Luca Turin, Mark Uncles, Dietmar Voelkle, Bernard Wentworth, Diederik Wiersma, Andy Yeatman, and Timothy Zwier.

God bless us, every one!

ROGER HIGHFIELD
Greenwich, Christmas 1997

THE PHYSICS OF CHRISTMAS

1

CHRISTMAS AND THE SCIENTIST

AN INTRODUCTION

There seems a magic in the very name of Christmas.
CHARLES DICKENS, SKETCHES BY BOZ

CHRISTMAS IS A TIME FOR THE CRUNCH OF SNOW, spiced wine, and tinseled trees. Christmas is a time for giving, meeting friends, and feasting. Christmas is a time for carols, family gatherings, gaudy greeting cards, and all the jollity of the seasonal spirit. Christmas is also a time for science.

Chemists are hard at work in the Christmas kitchen. Experts on thermodynamics have drafted equations to help us cook turkeys to perfection, scanners have scrutinized steaming plum puddings, and pharmacologists have traced the baroque metabolic pathways of the brain to explain why chocolates can be so addictive.

Meteorologists study every aspect of the snow cycle that provides a seasonal sprinkling, from the seeding of an ice crystal high in the sky to the traces of past Christmases buried deep in the snowpack.

Climatologists are plundering this record to help predict white Christmases far into the future. A handful are even concocting outlandish schemes to guarantee that each and every Christmas is white.

Psychologists tease out the hidden agenda of the Christmas card and what it reveals about our social status. The same goes for presents. The price, the nature of the gift, and even the way it is wrapped say a great deal about the giver and his or her relationship with the recipient. All the while, anthropologists hunt for the foundation of the celebration in pagan rituals that took place before the birth of Christ, during long winter nights when our ancestors feared that the sun would never return.

The origins of the holiday in the darkness of prehistory emphasize perhaps the most fundamental aspect of Christmas: everyone's invited. The seasonal message of hope and charity is a message for all — Christians, Jews, Hindus, Moslems, Buddhists, and, yes, even scientists and engineers.

I have been investigating the science of Christmas for more than a decade. When I first began to take an interest in the subject, I was unprepared for the breadth and depth of the insights that would eventually emerge. Take those flying reindeer, Santa's red and white color scheme, and his jolly disposition, for example. They are all probably linked to the use of a hallucinogenic toadstool in ancient rituals.

I can add that Santa was born with a genetic propensity to become obese and now suffers from diabetes. He does not live at the North Pole, preferring the warmth of an island off the coast of Turkey. There, panting at his side, you will find Rosie — not Rudolph — the reindeer.

I was at first puzzled by how Santa could fly in any weather, circle the globe on Christmas Eve, carry millions and millions of presents, and make all those rooftop landings with pinpoint accuracy. The answer lies in his unprecedented research resources and expertise across a range of fields, spanning genetic engineering, computing, nanotechnology, and quantum gravity.

My experience of writing this book undermines the idea that the materialist insights of science destroy our capacity to wonder, leaving the world a more boring and predictable place. For me, the very reverse is true. I can still remember the day when, as a child, I first became convinced that Santa did not exist. Now, by refracting the Santa myth through the prism of science, he seems more real than ever.

I believe that science and technology can even shed a little light on a deeper question: where did Christmas come from in the first place? Peel back the wallpaper of centuries, and you will find that the festival is an amalgam of influences — German, Dutch, English, American, and other traditions, both religious and pagan — that emerged over the millennia.

Even today, the traditional Christmas hoopla is far from a homogeneous phenomenon, taking place alongside Kwanzaa, an African-American harvest holiday, and the eight-day Jewish celebration of Hanukkah. Together they constitute *the* annual celebration.

Part of the reason winter festivities went global can be found 150 years ago, at the tail end of the Industrial Revolution. It was then that "Christ's Mass" (*Cristes maesse* in Old English), the church service that celebrates the birth of Jesus Christ, along with a wealth of other traditions, entered the scientific age of mass communications, transport, and other technologies.

This collision between ancient tradition and the age of

science and technology was particularly significant in Victorian Britain, where, during a single decade, there was a striking coincidence of events of significance for science, the annual celebrations, and this book.

The 1840s saw a dizzying rate of change in society due to efforts across a proliferating range of disciplines. In the world of science, there were Darwin's ideas on natural selection, Joule's work on thermodynamics, and Faraday's studies of magnetism, light, and electricity.

In the sister disciplines of engineering and technology, there were developments in factories, machine tools, and information technology. Babbage was hard at work on his difference engine, and a web of telegraph lines spread across the nation. All the while the old certainties seemed to have been squashed flat by the steam hammer, steamboat, and steam train. The resulting turmoil in society made the traditional Christmas message of charity more relevant than ever.

Emerging communications technologies, from speedy railways to the telegraph, paved the way for that message to be disseminated and homogenized for mass consumption, forging much of what we think of today as the traditional festivities.

The tumultuous 1840s also saw an important token of the rising influence of science: the birth of a specific label for the burgeoning army of individuals at work in this field. William Whewell, a polymath who was a Fellow of the Royal Society, coined the word *scientist* in earnest in his two-volume book *The Philosophy of the Inductive Sciences.* The word was of dubious legitimacy in philological terms, a hybrid of Latin and Greek, and was attacked (wrongly) as "an American barbarous trisyllable." But the pressure to put a name to this increasingly influential group was overwhelming.

That same decade saw the introduction to Britain of one component of the German Christmas that remains very much a part of the celebrations today. Queen Victoria and Prince Albert set up a Christmas tree for the first time in Windsor Castle in 1840. She recorded that this German custom quite affected dear Albert, who turned pale and had tears in his eyes! Eight years later they appeared beside the tree in the *Illustrated London News*, one of the magazines established that decade to exploit advances in illustration technology. This would become one of the most famous nineteenth-century Christmas scenes of all.

At the same time that *scientist* was born and Albert gazed upon his tree, an eminent and extraordinary individual, Henry Cole, decided to reduce the burden of writing Christmas greetings letters by taking advantage of another development he had had a hand in: the introduction of the penny post in 1840.

His invention, the first Christmas card, was published in 1843 and cost a shilling, the equivalent of a day's wages for a laborer. After two decades the price fell dramatically thanks to one of the technological innovations of the day, cheap color lithography, and Christmas cards entered the mass market.

Cole regarded the card as the folk art of the Industrial Revolution, and it ultimately became the greatest popularizer of now-standard Christmas iconography, with designs ranging from bizarre characters with pudding heads to mannequins in period costume, as well as the more conventional mistletoe, robins, holly, and fireside scenes. Not only were the cards printed on paper, but they were also gilded, frosted, and dressed with satin or fringed silk. Some were even made to squeak.

Through the evolution of one of the card's most familiar characters, it is possible, in the wake of the pioneering contributions of Cole, Prince Albert, and Whewell, to trace the influence of scientists, engineers, and technologists on our way of life. I am, of course, referring to the many depictions of that fat man with the white beard.

A silk-fringed card published in 1888 reveals how, by then, Santa had resorted to the latest communications technology to improve links with his market. The figure shown on the card seems to be engaged in what can only be described as a conference call, listening to the simultaneous demands for presents from an assortment of children. Only the previous decade, Alexander Graham Bell had patented the telephone that made it all possible.

By the 1890s Santa had decided to give up his sleigh and reindeer, preferring to haul his gifts around by "the new monstrosity from France," the automobile. As a result of the development of the internal combustion engine, the silent night, holy night now throbs to the sound of traffic. The stillness of the snowy landscape shown on so many Christmas cards is marred by the groan of the snowplow and the susurrus of chains on wheels. The search for the Bethlehem star is now obscured by a haze of photochemical smog.

Another newfangled device, the wireless, appears on one 1929 Christmas card, which features a Santa apparently mesmerized by the crackling message it is receiving over the ether: "You're in my Christmas circuit / And on the waves of thought / A Happy Christmas and New Year / To you is gladly brought." Radio would become the first mass medium to reinforce the tendency for Christmas to be a festival held behind closed doors.

When Santa reached for a cool soda pop in a Coca-Cola

advertisement that appeared during the Christmas season of 1937, he was again a technological pioneer. The source of his refreshment was a refrigerator, even though iceboxes were still being used by most American households that year.

Santa can now be found in cyberspace. The last time I checked, there were hundreds of Santa home pages for children's e-mail. Digitized images of Santa now scud about the web of international computer networks every Christmas.

One day these images may even supplant the traditional Christmas card. However, I believe that an e-mailed Santa, spouting digital "ho, hos" and seasonal greetings, would still honor the spirit in which Henry Cole first dreamed up the card — as a practical way to marry mass communications and art.

Cole would be amazed and gratified by the extent to which his little invention has caught on today. The significance of the 1840s does not end there, however. As Cole sent out his first cards, the greatest and most influential of all Christmas books made its first appearance in a crimson and gold binding.

A Christmas Carol was published by Chapman and Hall on December 19, 1843. By Christmas Eve it had sold six thousand copies, the most successful publication that season. Within two months eight pirated theatrical productions had been staged.

The genesis of this work of popular genius dates back to around 1840 and Dickens's correspondence with the philanthropist Lord Ashley. Dickens was horrified by the impact on society of the age of machines, notably the appalling conditions endured by children working in coal mines and factories. He started work on the book to make a sledgehammer blow against these evils of the industrial age.

One newspaper described the book as "sublime." Thackeray said that it was a "national benefit." Lord Jeffrey told Dickens that it had "prompted more positive acts of beneficence than can be traced to all the pulpits and confessionals in Christendom since Christmas 1842."

Thus the 1840s saw a striking convergence: the first scientist, the tree, the card, and the Christmas book to top all Christmas books. A century and a half later, science is still altering the very nature and fabric of the celebrations through the introduction of new technology, whether cloned Christmas trees, the Internet, or those infuriating cards that play carols over and over again.

And so on to the science of Christmas.

2

SANTA AND THOSE REINDEER

His eyes how they twinkled! His dimples how merry!
His cheeks were like roses, his nose like a cherry.
His droll little mouth was drawn up like a bow
And the beard on his chin was as white as the snow.
The stump of a pipe he held tight in his teeth,
And the smoke, it encircled his head like a wreath.
He had a broad face and a little round belly
That shook, when he laughed, like a bowl full of jelly.

CLEMENT CLARKE MOORE,
"A VISIT FROM ST. NICHOLAS"

WHERE DO YOU THINK SANTA CLAUS IS RIGHT now? Sitting with a glass of sherry in front of the glowing embers in a cozy wooden house while Arctic snow falls softly on his sleigh outside? Or maybe feeding the reindeer? Perhaps he has his maps out and is making adjustments to his route across the North Pole for Christmas Eve?

Not the *real* Santa. For the sake of accuracy, Christmas cards should show Santa in sunglasses, clad in red and white swimming trunks, and sipping a cool Coke next to a swimming pool. For the sake of completeness, a reindeer with a sunburned nose, called Rosie, should be panting nearby.

There is now evidence to suggest that Santa's abode lies not on the polar ice cap, but among Mediterranean olive groves on Gemiler, a tiny island off Turkey. It is there, historians believe, that St. Nicholas, a direct ancestor of Santa Claus, may have died.

Gemiler is well-known to tourists and has recently been the subject of a number of archaeological studies, most recently by the University of Osaka, and by a group of scholars including David Price-Williams, an archaeologist who lectures at London University. Though it is only half a mile long, it has at least five churches decorated with frescoes and mosaics and all the hallmarks of a major religious site — a holy city dedicated to St. Nicholas.

Medieval Venetian sailing instructions refer to Gemiler as the Island of San Nicolo. On a church door near the anchorage is a painting of "Osios Nikolaus" — St. Nicholas himself. The island also has a huge Byzantine ecclesiastical complex, with a magnificent 300-meter barrel-vaulted processional way. At other Byzantine sites processional roadways are often associated with monastic complexes dedicated to the veneration of major saints, but few ever reached the grandeur of the one at Gemiler.

Who Was Santa?

Legend suggests that St. Nicholas was born around A.D. 245 in the town of Patara, an important Byzantine port in Turkey, only a couple of hours' sail from Gemiler. When Nicholas was a young man, his father died, leaving a great fortune. Nicholas began anonymously giving away the money to the needy, especially to children. Eventually he became Bishop of Myra (the modern-day coastal town of Demre), at the southernmost tip of the Bey Daglari Mountains. (The name "Myra" is derived from that of the resin myrrh.) There he supposedly performed several miracles, including saving sailors from drowning and resurrecting three boys who had been killed by an evil butcher. It is the best-known of his miracles, however, that helps to wrap St. Nicholas into the legend of Santa Claus.

This miracle concerned a noble and his three daughters, who had fallen on hard times. The daughters had little chance of marriage, as their father could not pay their dowries, so they faced a life of prostitution. One night St. Nicholas, hearing of the girls' plight, threw a sack of gold through a window of the nobleman's shabby castle. The sack contained enough gold to provide for one daughter's marriage. The next night he tossed another sack of gold through the window for the second daughter. But on the third night the window was closed. Ever resourceful, St. Nicholas dropped the third sack of gold down the chimney. Townsfolk heard the story and began hanging stockings by the fireplace at night to collect any gold that might come their way, presumably — hence the tradition of the Christmas stocking and Santa's affinity for fireplaces.

St. Nicholas probably died sometime in the mid-fourth century. (One oft-quoted date is December 6, 343.) The earliest Byzantine portraits show him with a long white beard, and when the reformed church spread throughout Europe, he became linked with Christmas because his feast day is 6 December. His fame was widespread by the sixth century — a possible explanation for the huge settlement on Gemiler.

But just after 650, this place of veneration was disbanded. The Islamic governor of Syria launched a fleet to challenge Byzantine sea power in the Mediterranean. He quickly destroyed the settlements on Cyprus, followed by those on Rhodes and Cos. Gemiler was abandoned. The site lay forgotten and forlorn — the lost sacred city of St. Nicolas. Today St. Nicholas remains one of the most popular Christian saints and is known as the patron of children, sailors, teachers, students, and merchants.

There are many and varied explanations of how St. Nicholas evolved into the character we know. All that can be said with certainty is that Santa's roots lie in folk customs and beliefs from a sackful of sources. These include the British Father Christmas, the French Père Noël, the Dutch Sinterklaas, the Danish Jules-Missen, and even the Romanian Mos Craicun.

The Protestant church also influenced the evolution of this icon. When Martin Luther objected to the practice of gifts being given to children in the name of a Catholic saint, Nicholas was joined during the Reformation by a child, the Christkindlein. This would mutate back into the Father Christmas figure Kriss Kringle in English-speaking society.

Then the Christkindlein was joined by a dwarfish, dark-faced companion, often a frightening figure, known variously as Krampus, Pelzebock, Pelznickel (Nicholas in furs), Hans Muff, Bartel, or Gumphinkel. There were also female equivalents — Berchtel, Buzebergt, and Budelfrau. Most commonly the companion was called Knecht Ruprecht and carried a bundle of switches to mete out punishment to naughty children.

The Dutch are often credited with transforming the saint into the character we know today. Their custom of giving presents to children on the Day of St. Nicholas was brought to America by early Dutch settlers of New Amsterdam (renamed New York when the British took over the colony). There Sinterklaas, the colloquial Dutch for St. Nicholas, evolved into Santa Claus.

Sinterklaas was traditionally depicted with a broad-brimmed hat and a pipe, and his long churchly robe was replaced with short breeches. By the beginning of the nineteenth century, the various traditions started to mingle, so that in 1809, for instance, the American writer Washington Irving wrote of a jolly, chubby fellow riding in a wagon over treetops.

There is another, quite different way to trace the evolution of the modern Santa. His development could be viewed in terms of how brains have been parasitized through the ages by entities that evolved to thrive in just such a niche. These are *memes* (loosely speaking, units of cultural transmission), a term coined by the biologist Richard Dawkins to show that ideas replicate rather like genes do. Examples include tunes, catchphrases, innovative concepts, clothes, fashions, and, of course, Santa Claus, Father Christmas, and the rest.

Genes are carried by organisms in which they produce effects (skin color, blood type, and so on) that make each of us individual. Memes are carried by meme vehicles — poems, books, sayings, and so on — bearing an idea that will distract us, burden our memories, and coerce children to be well behaved in the frantic run up to Christmas Day. Otherwise, as the memes warn, Santa will not deliver any presents. As one sociologist puts it, "Parents use the belief in Santa Claus to control children, to induce children to defer demands for gratification to Christmas, and to make it appear that Santa, not the parents, causes the deprivation of children."

Modern Santa and Meaning

It would be a mistake to describe today's Santa as a simple amalgam and evolutionary endpoint of his rich mixture of ancestors. For one thing, many versions still exist. In different regions of Germany St. Nick is known by various names, including Klaasbuur, Burklaas, Rauklas, Bullerklaas, and Sunnercla. In eastern Germany, where the Santa figure remains more connected with his pagan past, he is called Ash Man, Shaggy Goat, or Rider. There is also the Weihnachtsmann, a Father Christmas–like figure who is depicted as tired and stooped from toiling through the dark winter night with his heavy burden of toys.

Another blow against the Santa-as-amalgam model has been struck by anthropologists. They have set to work on the most ubiquitous form of the modern Santa and declared him to be more than the sum of his European influences — indeed they see him as distinctly American. They highlight five key differences between the Santa of today and his ancestors: (1) Santa lacks the religious baggage of his predecessors; (2) he is, by the standards of Knecht Ruprecht, a bit boring; (3) he has turned into a softhearted liberal with no stomach for the punishment meted out by the likes of Sinterklaas and Knecht Ruprecht; (4) this mythical figure is more tangible than his predecessors, thanks to appearances in films, TV shows, and department stores (even in Japan); and (5) he spends much more than his central European forebears, preferring to give Nintendo video games rather than nuts, for example.

The distinguished anthropologist Claude Lévi-Strauss has provided a wonderful pen portrait of this Christmas icon: "Father Christmas is dressed in scarlet: he is a king. His white beard, his furs and his boots, the sleigh in which he travels evoke winter. He is called 'Father' and he is an old man, thus he incarnates the benevolent form of the authority of the ancients."

Importantly, says Lévi-Strauss, children believe in him, paying homage to him with letters and prayers, while adults do not: "Father Christmas thus first of all expresses the difference in status between little children on the one hand, and adolescents and adults on the other. In this sense he is linked to a vast array of beliefs and practices which anthropologists have studied in many societies to try to understand rites of passage and initiation."

Sociologists have also been toiling away to reveal what we mean by Santa. Warren Hagstrom of the University of Wisconsin, Madison, couches his analysis in terms of either positivism or Clauseology. For the positivist (nineteenth-century version), "belief in Santa Claus is defined as erroneous; and the problem of the positivist is to discover how such erroneous beliefs arise. The positivist, arguing that all beliefs arise by inference from experiences, finds the meaning of Santa in false inferences from actual experiences."

The naturism of the German-born British philologist Max Müller is a variety of positivism that finds the origins of figures like Santa in natural phenomena, says Hagstrom. Children, like primitive people, often personalize the forces of nature. "While small children may find it difficult to conceptualize the winter solstice, they find it easy to conceptualize Santa Claus. (Ask any child questions about the two phenomena.)"

The Clauseologist position, Hagstrom explains, is that Santa Claus exists but that his essential nature ("meaning") cannot be empirically ascertained. "The empirical phenomena associated with Santa are likely to be illusory and deceptive. It is instead necessary to rely on nonempirical methods of investigation, of which there are two types: inner experience and revealed sources. I cannot report here my inner experiences of Santa Claus, since it has been so long since I've had any genuine experiences of this type." This piece of whimsy, published in *American Sociologist,* goes on to say that one of the major problems facing Clauseologists is collecting authentic revealed sources.

Fortunately, Hagstrom accepts works like "A Visit from St. Nicholas" as part of the canon. This Christmas poem marks perhaps the most important single blueprint for modern Santa. It was written by Clement Clarke Moore, a professor at the General Theological Seminary in New York. A classical scholar and poet, Moore had translated Juvenal and other Roman poets into English verse and turned his hand to poetry in the romantic style. He was familiar with the folklore of the Dutch, German, and Scandinavian immigrants who had settled in the northern United States, including the Dutch tradition of Sinterklaas (which by then was widely observed on December 24 and 25) and the Teutonic and Norse notions of a jovial but somewhat impish figure who presided over the pagan midwinter festivities. In 1822 he synthesized the lot into a figure who stars in his poem "A Visit from St. Nicholas."

That December Moore read the verses aloud to his children. A visitor to his home was so impressed that he had the poem published the following year in the *Troy Sentinel* in upstate New York. The poem gave us these oft-quoted lines: "'Twas the night before Christmas, when all through the house / Not a creature was stirring — not even a mouse." In dozens of rhyming couplets, often derided today as doggerel, he described a plump, pipe-smoking Santa who traveled from the north in a sleigh drawn by tiny flying reindeer with "dainty hooves." This St. Nicholas also had a belly "that shook . . . like a bowl full of jelly" and a beard that was "white as the snow." That much sounds very familiar. However, he was "dressed all in fur, from his head to his foot," which is more reminiscent of Pelznickel than of a latter-day Santa.

Alas, St. Nicholas probably did not celebrate Christmas and probably never saw, or even knew about, reindeer. In Dutch legends Sinterklaas travels on a gray horse and wears bishop's robes. It is not clear when, if ever, Moore saw a sleigh drawn by reindeer, let alone the beasts in the wild, though he may have been acquainted with a Finnish legend concerning "Old Man Winter," who drives his reindeer down from the mountains, bringing snow with him.

Further evolution in the image of Santa occurred when he was depicted as a pear-shaped, jolly character with a flowing white beard in drawings by Thomas Nast in *Harper's Weekly* between 1863 and 1886. The break with his religious past was by then clear: Nast's Santa was reminiscent of his drawings of a drunken Bacchus and the corpulent plutocrat William "Boss" Tweed. Nast himself admitted that he was also inspired by the furs of the Astors when he designed Santa's fur-trimmed garb.

When it comes to the kind of Santa that we see stalking shopping malls and TV today — the jolly, fat figure clad in red and white — a leading manufacturer of carbonated beverages claims the credit for that archetype. A year or two ago, Coca-Cola even had the cheek to celebrate Santa's sixty-fifth birthday.

Before 1931, the company says, Santa Claus appeared in many different guises, from a green elf to a somber St. Nicholas and even a gaunt figure dressed in animal skins. That year, so the publicity goes, Coca-Cola commissioned a young Swedish artist, Haddon Sundblom, to give the icon a makeover.

From 1931 on, Sundblom created at least one Santa picture annually. His St. Nicholas wore an ample red coat trimmed in white and held in place with a thick leather belt, and he was depicted in various seasonal scenes. A hat, also trimmed in white, appeared in 1934. Sundblom removed Santa's pipe, which can be seen in Nast's creation, and gave him a bottle of Coke.

Through a succession of poses — with children, reindeer, sacks of toys, or letters — he was never without his fizzy drink. With the billowing beard, expansive girth, and rosy cheeks, he would gaze intently at his bottle or grasp it heartily, ready for that "pause that refreshes."

Santa: The Hallucinogenic Connection

A rival suggestion for the origins of much of Santa's paraphernalia — his red and white color scheme, those flying reindeer, and so on — is much more fun, less commercial, more scientific, and somehow more appealing than Coca-Cola's version, because it is so politically incorrect.

Patrick Harding of Sheffield University in England argues that the trappings of the traditional Christmas experience owe a great deal to what is probably the most important mushroom in history: fly agaric *(Amanita muscaria),* the recreational and ritualistic drug of choice in parts of northern Europe before vodka was imported from the East. Each December this mycologist dresses up as Santa and drags a sleigh behind him to deliver seasonal lectures on the toadstool. The garb helps Harding drive home his point, for Santa's robes without doubt honor the red-and-white-dot color scheme of this potent mind-altering mushroom.

Commonly found in northern Europe, North America, and New Zealand, fly agaric is fairly poisonous, being a relative of the more lethal death cap *(Amanita phalloides)* and destroying angel *(Amanita virosa).* The hallucinogenic principles of fly agaric are due to the presence of the chemicals ibotenic acid and muscimol, according to the International Mycological Institute at Egham, Surrey, England. Ibotenic acid is present only in fresh mushrooms. On drying, it turns into muscimol, which is ten times more potent. In Lapp societies, the village holy man, or shaman, took his mushrooms dried — with good reason.

The shaman knew how to prepare the mushroom, removing the more potent toxins so that it was safe enough to eat. During a mushroom-induced trance, he would start to twitch and sweat. His soul was thought to leave the body as an animal and fly to the otherworld to communicate with the spirits. The spirits would, the shaman hoped, help him to deal with pressing problems, such as an outbreak of sickness in the village. With luck, after his hallucinatory flight across the skies, he would return bearing the gifts of medical knowledge from the gods.

Santa's jolly "Ho, ho, ho" is the euphoric laugh of someone who has indulged in the mushroom. Harding adds that the big man's fondness for popping down chimneys is an echo of how the shaman would drop into a yurt, an ancient tentlike dwelling made of birch and reindeer hide. "The 'door' and the chimney of the yurt were the same, and the most significant person coming down the chimney would have been a shaman coming to heal a sick person."

Harding uses the shaman's urine to link reindeer to the myth. For one thing, reindeer were uncommonly fond of drinking human urine that contained muscimol. The hoi polloi from the village also were partial to mind-expanding yellow snow, because the potency of the muscimol was not greatly weakened — although it was probably safer — once it had passed through the shaman. "There is evidence of the drug passing through five or six people and still being effective," Harding says. "This is almost certainly the derivation of the phrase 'to get pissed,' which has nothing to do with alcohol. It predates inebriation by alcohol by several thousand years."

Such was the intensity of the drug-induced experience that it is hardly surprising that the Christmas legend includes flying reindeer. Witches soar for related reasons: a witch who wanted to "fly" to a sabbat, or orgiastic ceremony, would anoint a staff with specially prepared oils containing psychoactive matter, probably from toad skins, and then apply it to vaginal membranes.

References to flying can be found in more recent applications of the mushroom. St. Catherine of Genoa (1447–1510) used fly agaric to soar to the heights of religious ecstasy, according to Daniele Piomelli of the Unité de Neurobiologie et Pharmacologie de I'Inserm in Paris. An account of the life of St. Catherine describes the use of ground agaric, so that God "infused such suavity and divine sweetness in her heart that both soul and body were so full as to make her unable to stand."

In Victorian times travelers returned with intriguing tales

of the use of fly agaric by people in Siberia, Lapland, and other areas in the northern latitudes. One of the first was reported by the mycologist Mordecai Cooke, who mentioned the recycling of urine rich in muscimol in his *A Plain and Easy Account of British Fungi* (1862). Harding points out that Cooke was a friend of Charles Dodgson (Lewis Carroll), the author of the fantastic children's story *Alice's Adventures in Wonderland* (1865). "Almost certainly, this is the source of the episode in Alice where she eats the mushroom, where one side makes her grow very tall and the other very small," Harding says. "This inability to judge size — macropsia — is one of the effects of fly agaric."

Rudolph the Red-Nosed Reindeer

Long before 1949, when that perennially popular Christmas hit "Rudolph the Red-Nosed Reindeer" was launched, the myth of the reindeer was already well established. English texts from the Renaissance mention the display of antlers during Christmas dances centuries before any belief in Father Christmas, much less the development of his legend.

Rudolph himself first appeared in an illustrated booklet written by Robert May in 1939 for the Montgomery Ward department stores to hand out to children at Christmas, and was used as the theme for the popular song written by Johnny Marks a decade later. It was first performed by Gene Autry, the "Singing Cowboy."

One commonly held view is that Rudolph's nose was red due to a cold. Others claim that the song has saddled Rudolph with the red-nose slur — the implication being that while Santa consumes the milk and cookies left out for him, Rudolph helps himself to the strong stuff. The unexpected triumph of the drunken, inefficient Rudolph over his sober companions chimes with the relaxation of social conventions that has long taken place during winter festivals.

Recent research conducted in Norway, however, offers a more convincing explanation. Unfortunately for Rudolph, reindeer noses provide a welcoming environment for bugs. They have elaborately folded turbinal bones covered with blood-rich membranes, which warm the air as they breathe in and cool it as they breathe out, thereby reducing the loss of both heat and water. (Even when there are icicles and frost on Santa's beard, his faithful reindeer have dry muzzles.) Odd Halvorsen of the University of Oslo suggested some years ago in the journal *Parasitology Today* that the "celebrated discoloration" of Rudolph's nose is probably due to a parasitic infection of his respiratory system. Even today, he is awed by the response that followed this revelation. "This paper brought me more fame than anything else I have published," he admits.

Despite living in such chilly conditions, reindeer not only share many of the same parasites that plague other ruminants, such as the warble fly, but also are preyed upon by around twenty different parasites that are specific to them. The pentastomid *Linguatula arctica,* one of a group of creatures called tongue worms, can be found in reindeer sinuses; larvae of the fly *Cephenemyia trompe* wriggle in the nasal cavity; and nematodes of the genus *Dictoycaulus* squirm in the lungs, as do vast numbers of *Elaphostrongylus rangiferi* larvae. "We have not been able to quantify the combined effects of these parasites, but it is no wonder that poor Rudolph, burdened as he is by parasites, gets a red nose when he is forced to pull along an extra burden like Santa Claus," Halvorsen notes.

Rudolph notwithstanding, it remains something of a puzzle why reindeer are so embedded in modern Christmas culture. They were only one among many kinds of grazing and browsing mammals that once roamed the forests and plains of Europe, northern Asia, and North America. Indeed, ancient reindeer remains suggest that they ranged as far south as Spain and Italy.

They had been established for around a million years by the time humans came on the scene. People hunted reindeer, along with bison, mammoths, wild horses, and many smaller mammals. Reindeer meat is delicious, the fur is light and warm, and the antlers and bones are handy for making tools and ornaments. No wonder the beasts are featured in cave art and rock carvings, such as one found in Sagelva, Norway, that dates back to 2000 B.C.

But reindeer are badly misrepresented during Christmas festivities, according to Caroline Pond of the Open University in England. Pond is a biologist who has studied reindeer with other biologists from the University of Tromsö, Norway. Take the depictions of the beasts typically found on cards, for example. True, reindeer are the only deer species for which both sexes have antlers — bone, often branched, that is covered with a thin layer of skin, or "velvet," rich in blood vessels. But the males actually lose their crowning glory around the time that the holiday is celebrated. The reason has to do with sex.

Antlers of mature male reindeer are usually larger than those of females, with the most impressive found in caribou and Norwegian reindeer. They probably evolved as a secondary characteristic of males under sexual selection: they depend on the sex hormone testosterone; are larger, more elaborate, and heavier in older males; and are at their biggest during the breeding season, when they are essential for ritual combat and fighting. Afterward, the males are "rutted out" (even Rudolph), exhausted by the loss of body weight and fat reserves. It comes as no surprise that studies have found that male reindeer suffer greater mortality than females.

Changes in the concentration of sex hormones promote bone reabsorption at the base of the antlers in adult males. Eventually the antlers fall off, and there is a delay of up to four months before new ones grow in the spring. Perhaps the inaccurate depictions of Rudolph sporting his antlers wish to deny this seamy side of reindeer life.

The reindeer Lapps, or Sami, an ethnic group living in northern parts of Sweden, Norway, Finland, and Russia, acknowledge this link between virility and antlers by selling powdered reindeer "horn" to the Japanese with the claim that it increases potency. The Sami are unusually virile, but the reason, according to a study by Ilpo Huhtaniemi of Finland's Turku University, is not due to this horny folk medicine but to a genetic mutation.

The mutation is found in 40 percent of Sami men (compared with 25 percent of other men in Finland and 20 percent of Swedes) and apparently maintains a high level of testosterone in older men. It seems that the farther south you go, the lower the incidence of the mutation. "The frequency of the mutation is 15 percent in men from southern Europe, 10 percent in Asian men, and 5 percent in American Indian men," Huhtaniemi says.

The very fact that Christmas card artists show Rudolph with his antlers in place may underscore another unfortunate fact, one drawn to my attention by Odd Halvorsen: the Sami mostly use castrated male reindeer to pull or carry loads.

Without their equipment, males have an abnormal antler cycle, so they keep their headgear longer than functional males. To keep his antlers for the sake of the Christmas card, Rudolph would have had to be castrated. "This introduces another sad aspect to the story," Halvorsen says.

The more we know about reindeer, the worse the problems faced by card illustrators become. While the males are squandering their energy on sex and violence, the females are piling on fat, Caroline Pond notes. By the time Christmas arrives, the only adult reindeer with antlers and enough energy to drag around a sleigh full of presents are females. That is why Marks's song should have been about Rosie the Red-Nosed Reindeer.

Reindeer are well adapted to living in a snowy landscape, though one that is more barren than the kind found on Christmas cards. In winter they dig through snow to feed on the plants underneath. Fine, powdery snow is easy enough for them to handle, but if the snow is too deep or too hard, feeding becomes difficult. Snow that melts and refreezes to form a crust of ice can be so firm that the reindeer cannot dig through it to reach the food underneath.

Other reindeer behavior is also misleadingly depicted on Christmas cards. The animals' fur is an efficient insulator; outer hairs are long and hollow, supporting a fine, dense undercoat. Together they trap a layer of warm air. Insulation is so effective that snow does not melt on the backs or heads of reindeer. Rudolph, Dasher, Prancer, and the rest of the crew are so well adapted to the cold that they would probably find loafing around chimneys and firesides with Santa too warm to be comfortable.

3

THE FLAME AND TREE

I have been looking on, this evening, at a merry company of children assembled round that pretty German toy, a Christmas Tree. The tree was planted in the middle of a great round table, and towered high above their heads. It was brilliantly lighted by a multitude of little tapers; and everywhere sparkled and glittered with bright objects.

CHARLES DICKENS, A CHRISTMAS TREE

A TREE FESTOONED WITH FLICKERING CANDLES can be found on many Christmas cards, and for good reason. The evergreen and the candle celebrate the same thing: life-giving sunlight, an ancient symbol that dates back long before Prince Albert introduced the tree to Britain or Martin Luther supposedly first bedecked a tree with candles in the sixteenth century to remind children of the heavens from which Christ descended.

Our ancestors held winter festivities to usher in the annual return of sunlight, warmth, and fertility with rituals involving evergreens, which seemed to defy the cold winter months, and the yellow light of a living flame. Today Christmas and other seasonal celebrations, such as Kwanzaa and Hanukkah, are united by this latter symbol of the rebirth of the sun's life-giving energy.

We can gaze deeply into the workings of the living world by studying what happens when we light a candle. The resulting flame marks the last step in an extraordinary series of physical and chemical processes that first capture sunlight to forge chemical bonds in wick and wax, then snap them to release the long-pent-up light.

The most important of the processes is photosynthesis, a word derived from *photo,* meaning "light," and *synthesis,* meaning "the production of something." Photosynthesis drives the living economy that thrives on the surface of our planet. Each year green plants, including Christmas trees, harness the energy of sunlight to pluck 100 trillion kilograms of carbon dioxide from the sky, then combine it with hydrogen from water to build carbohydrates — their food — and to release oxygen.

One acre of Christmas trees can produce the daily oxygen requirement for eighteen people. In the United States there are approximately one million acres of growing Christmas trees; that means around eighteen million people each day are supplied with the oxygen generated as the trees harvest sunlight.

What is so beautiful about the act of lighting a candle on a Christmas tree is that it honors the cycles that turn within and without living things. The chemical energy generated by photosynthesis in plants is passed up the food chain, for instance, to grazing cattle and then on to a tallow candle. When the candle is lit at the gloomiest time of year, it releases this "cryptic sunlight" and returns the complex fat, or wax, molecules to the form in which the plants found them — water and a hot breath of carbon dioxide that can again be incorporated into living things.

The Christmas Tree

Like so many aspects of holiday celebrations, the roots of this symbol stretch back to prehistoric times, when ancient people became fascinated by how some trees and plants continued to thrive among the dead branches of a forest in winter. To a primitive mind in a deciduous world, an evergreen suggested permanence and a magical ability to endure with little help from the sun.

The ancient Egyptians brought green palm branches into their homes on the shortest day of the year, in December, as a symbol of life's triumph over death. This symbolism is also apparent in the Roman festival of Saturnalia, when buildings were decorated with evergreen branches of holly, pine, and ivy in honor of Saturnus, the god of agriculture.

Holly, ivy, and mistletoe are not only green but also bear recognizable fruit during the winter, again paying little heed to the elusive sun and keeping alive the hope that a fruitful year is to come. This triumph of fertility over the elements is echoed in the English legend of the Glastonbury thorn, planted by Joseph of Arimathea. The legend goes that soon after the death of Christ, Joseph went to Britain to spread the message of Christianity. Being tired from his journey, he lay down to rest and pushed his staff into the ground beside him. When he awoke, he found that the staff had taken root. The resultant bush was the Glastonbury thorn, which flowered each year on Christmas Day.

When discussing the prehistory of the Christmas tree,

one also should take account of the Yule log, which was made from timber big enough to burn throughout the longest winter night, again to help usher in the return of the sun. The resulting ashes supposedly had the power to protect a house against lightning, to cure maladies, and to fertilize fields.

The use of the fir as a Christmas tree started in ancient times in the Black Forest in southwestern Germany, where evergreens and small trees were part of the winter solstice festival of pagan tribes. But it is not known when the fir, or *Tannenbaum,* was adopted in other parts of the country. Legend has it that Martin Luther (1483–1546) was so moved by the brightness of the millions of stars on a winter's night that he set candles on his tree to simulate the effect. However, the idea of a decorated tree dates from much earlier, being reminiscent of tree-dressing rituals, which can be found from Russia to India in forms such as Yggdrasil, the Nordic tree of life; the Indian *bodhi* tree; and Eden's tree of knowledge. In various pagan rituals, a tree was decorated to encourage the tree spirits to return to the forest so that it would sprout again, which of course it did every spring.

Decorated trees are recorded in 1605 in homes in Strasbourg, then part of Germany. By 1796 we have the first illustration featuring a candlelit Christmas tree, showing Christmas Eve at Wandsbek Castle, near Hamburg.

In time the tree was brought to America by German settlers and by Hessian mercenaries paid by the British to fight in the Revolutionary War. In 1804, two hundred years after the first sighting of decorated trees in Germany, soldiers stationed at Fort Dearborn (now Chicago) were seen to haul trees from the surrounding woods to their barracks at Christmas.

Charles Minnegrode introduced the custom of decorating trees to Williamsburg, Virginia, in 1842. Almost a decade later, Mark Carr hauled two ox sleds loaded with trees from the Catskills to the streets of New York City and opened the first retail lot in the United States. By the end of the nineteenth century, the tree was so much a part of seasonal festivities that a description of Christmas at the White House refers to "an old-fashioned Christmas tree." Today around forty million American families celebrate the holidays with the fragrance and beauty of trees.

In Britain the custom of decorating Christmas trees was known to the socially aware from the middle of the eighteenth century because of the German background of the Hanoverian monarchs. It was not, however, until Queen Victoria and Prince Albert set up their tree for the first time in Windsor Castle in 1840 that this symbol was firmly fixed at the heart of the Victorian Christmas. Eight years later Victoria and her family appeared beside the tree in the *Illustrated London News,* marking one of the most famous nineteenth-century Christmas scenes of all.

Today the symbolism of the flame and tree burns as brightly as ever. In Oberammergau, Germany, trees planted on graves are decorated with glowing tapers to allow the dead to take part in Christmas celebrations.

Why Christmas Trees Are Evergreen

Leaves and needles are highly efficient solar cells, and to understand why a Christmas tree hangs on to this power supply during the dark winter months while others shed it, we need to understand a little more about the biochemical machinery that turns within their leaves.

The needles, fronds, and palms of green tissue provide plants with energy by harnessing sunlight through the process of photosynthesis. This occurs inside structures called chloroplasts, contained within leaf cells. There pigment molecules capture light energy. Chlorophyll is the most important pigment, absorbing red and blue light and allowing green wavelengths to reach the eye.

Winter sees less daylight combined with a sun that hangs much lower in the sky. That means less sunlight, a decline in the amount of photosynthesis, and thus less energy to sustain an individual tree. Moreover, temperatures become cold enough to threaten cell damage. Trees have adapted by resting through this period and living off excess food stored during the summer months. This process is not triggered solely by cold weather, even though chemical reactions within the leaf do slow in response to lower temperatures. The signal for dormancy is the decline in the hours of sunlight and the longer periods of darkness.

Trees have evolved different strategies to deal with the annual decline in sunlight. In general, they can be categorized as either deciduous (Latin for "to fall") or evergreen, a group that includes holly and conifers such as pines, junipers, and firs. Both types of trees cease growth in winter, but whereas deciduous trees shed all their leaves for this event, evergreens do not. Various factors govern when and why evergreens shed their needles. For example, some needles are lost if conifers are shadowed for any appreciable time or if deprived of water.

One might think that there is a simple reason why deciduous trees are stripped during winter: their broad leaves are less able to cope with winds and other vagaries of winter weather when compared with the aerodynamic needles of firs. To some extent this is true. Nonetheless, the loss of leaves is a deliberate affair that may be hastened by windy weather but is not dependent on it.

According to Bill Proebsting of Oregon State University, an elaborate cellular mechanism governs the process by which leaves and trees part company. At the base of each leaf is a special layer of cells called the abscission zone. When the time comes to shed the leaf, the cells in this layer begin to swell, slowing the transport of materials between the leaf and the tree. Once the abscission zone has been blocked, enzymes begin to break down the matrix holding the zone's cells together. A tear line is formed at the top of the zone and progresses downward, and eventually the leaf is blown away or falls off. Once the leaf falls, the stem side of the abscission zone forms a protective layer to seal the wound, preventing water from evaporating and bugs from getting in.

Peter Davies, a professor of plant physiology at Cornell University, says that several factors influence why Christmas trees and other conifers evolved to hang on to their needles in the winter rather than shed them all in one go. Dressed in their leaves, evergreens can carry out photosynthesis to take advantage of the light during the occasional days of winter sunshine. As a result, the leaves must be coated with a resin or wax to prevent moisture loss, and they must possess modified biochemical machinery to prevent cell damage caused by winter cold. This means that it takes evergreens more energy to make and maintain leaves as compared to trees that have thin leaves that are too delicate to withstand low temperatures. In effect, deciduous trees have decided not to invest any more resources in their doomed solar panels.

Within each leaf of a deciduous tree the green chlorophyll that captures sunlight and the biochemical machinery that converts sunlight into food gradually degrade. As the green pigment fades, it reveals other pigments, notably the yellow and orange carotenoids that provide vivid autumn colors, and reduces the ability of the leaf to harness sunlight. By shedding leaves, deciduous trees can maintain their energy efficiency at a slightly higher level.

Once the plant has withdrawn any useful nutrients and stored them in the cells of the trunk for use the next spring, leaf loss also can help rid the plant of the by-products that build up in leaves. One suggestion is that leaves are "excretophores" and that the shedding of a yellowed leaf is the equivalent of going to the toilet. This explanation adds an entirely new dimension to the ancient obsession with evergreens.

Supertree

The Christmas tree is under threat from a vulgar impostor: plastic trees that shed no needles, show no flaws, and can be used year after year. This is taking the deep-rooted obsession with the evergreen a step too far.

All is not lost, however. Thanks to the science of cloning, the real thing is staging a comeback. Various teams are developing methods of mass-producing thousands of copies of an individual high-quality Christmas tree.

The problem with the old-fashioned varieties of Christmas trees is that their seeds contain genetic variability, not only spelling imperfection for the consumer but also creating a headache for the grower when it comes to harvesting trees of many shapes, girths, and heights. To produce genetically homogeneous tree seeds, the traditional way would take about seven generations of breeding. Each generation takes fifteen to twenty years to mature, so it is just not feasible to obtain genetically uniform seeds this way. The alternative is to use cloning to create inbred lines of trees.

Dan Keathley's group at Michigan State University is focusing its cloning efforts on Douglas fir and Scotch pine. The first step in cloning, says Keathley, is to find "individuals that are really select. What we are talking about is the one tree in ten thousand." This flawless tree should have the following properties: a straight trunk that easily slips into a stand, the strength to hold lots of ornaments and tinsel, thick needles, good color, limbs angling upward at forty-five degrees, a uniform conical shape that tapers upward at thirty-five to forty-five degrees, and good needle retention. Growers want these properties to be coupled with disease and insect resistance and rapid growth. Mass production of such supertrees should cut costs, he says.

Once the supertree has been selected, a piece is cloned to produce a field of identical trees by one of two basic methods. In one process, the *tissue culture technique,* the researchers take a bud or seed from a tree, sterilize it to kill any fungi or bacteria present, and place it in a nutrient medium that promotes growth into a callus — a clump of randomly dividing cells. The callus can be broken into many pieces, each of which puts out shoots. In turn the shoots are placed in a medium that causes roots to grow using hormones. From there each shoot becomes a whole tree again.

The preferred method, *micropropagation,* involves taking a bud from a parent tree. Shoots can be produced from the bud with hormones and then grown into a tree. This avoids going through unorganized callus tissue, where chromosomal problems arise. "We are working to develop orchards for all the commercially important Christmas tree species," Keathley explains.

Hundreds of cloned Virginia pines have already been grown by another team at the Texas Agricultural Experiment Station. "The cloned tree is the future for Christmas trees in the South," says Don Kachtik, a grower whose test-tube-to-tree-farm pines are among the first in the state ready to be cut for the holidays. "We have only one type of tree that grows well in southeast Texas, the Virginia pine, so we need a good one." Like Keathley, the Texas team has found that tree farmers can save a great deal of time producing better trees by breeding good ones, then duplicating them by cloning. The traditional Virginia pine seedling costs less than 7 cents, compared to the current cost of 75 cents to $1.50 for a cloned tree ready to plant. But team member Ron Newton says that the expense of cloned tree plantlets is expected to decrease when the trees are mass-produced. Once they are a common feature of the holidays, it is unlikely that cloned trees, especially those improved genetically to resist insects and disease, will mean a higher bill for consumers.

According to Mike Walterscheidt, forest specialist for the Texas Agricultural Extension Service, "Growers hope that with cloning, they will be able to sell 95 percent of the trees they plant. Right now, only 60 to 70 percent are sold, because the others are ugly or die from disease or something."

Another opportunity presented by modern science is genetic engineering, where genes are inserted into a tree to introduce novel and desirable traits. Krishna Podila of Michigan Technological University argues that there are many reasons to engineer a supertree. The wood pulp industry, for example, would be able to produce, among other things, cheaper

Christmas cards. Genes that make the trees resistant to disease are also a goal of genetic engineers, since pine and spruce fungal diseases cause many losses in tree plantations.

Resistance to insects is another aim, since Christmas tree plantations are routinely attacked by soilborne insect larvae such as white grubs. Better needle retention might be achievable through the manipulation of genes involved in triggering senescence, the deterioration that accompanies aging. Scientists such as Podila also would like to boost the rate of growth of Christmas trees, which would reduce the turnaround time on tree plantations.

Podila's laboratory was one of the first to genetically engineer a conifer (a larch), and some progress in generating genetically engineered spruce and radiata pine trees has been reported by Dave Ellis in Madison, Wisconsin, and Christian Walter in New Zealand, respectively. Alas, conifer trees have proved extremely recalcitrant to manipulation through genetic engineering.

Those who fear that one day a "Frankentree" may run amok need not worry, it is claimed. For one thing, transgenic trees containing value-added genes to promote fast growth, herbicide resistance, and disease resistance will also be designed to be sterile, protecting both the environment and the interests of forest product companies. Such trees would be harvested earlier than conventional trees, which would prevent the possibility of genetic pollution caused by the spread of their genes.

At the moment, however, commercially produced, genetically engineered trees are but a glint in the eyes of biotechnologists. But it is only a matter of time until they become a reality.

The Candle at Christmas

No one knows who actually invented the candle. We do know that candles are the descendants of lamps that burned liquid fuel and date back at least thirty thousand years to the days when prehistoric people used them while daubing pictures deep inside caves.

Until the risk of deadly blazes became apparent and electricity provided an alternative, the flickering light of the candle was the standard method for illuminating Christmas trees. Like so many Christmas traditions, the use of candles can be traced to Germany, where the season begins with the start of Advent, four Sundays before Christmas Day. This period is marked by the Advent wreath, which originated with the Lutherans. The wreath consists of a circle of evergreen branches entwined with red ribbon and holding four candles, three purple and one rose or pink. One candle is lit to mark the passing of each week of Advent. Often a fifth white candle is placed at the wreath's center. This is the Christ Candle, symbolizing Christ's birth, and it is lit on Christmas Eve (*Heiligabend*) or Christmas Day. When it is bleak and cold outside, the wreath reminds us that there is life and freshness in the middle of winter.

Light continues to play an important role in many seasonal celebrations. Christians in China celebrate by using paper lanterns to turn Christmas trees into "Trees of Light." In parts of India, clay oil-burning lamps are used. In Sweden, people participate in Lucia processions, in which they give thanks to the Queen of Light for bringing hope during the darkest time of the year.

A candleholder called a Kinara is featured in the nonreligious holiday Kwanzaa, which was introduced after the Watts riots in the 1960s. Candles also are lit on the menorah or Hanukkiya, a seven- or nine-branched candelabrum, during Hanukkah, the Jewish festival of lights celebrating the victory of Judas Maccabees over the Syrian tyrant Antiochus more than two thousand years ago.

Why Does a Candle Burn?

Every Christmas some of the leading scientists of the day attempt to enlighten an audience of children during a series of lectures held at the Royal Institution in London. The great pioneer of electricity and magnetism, Michael Faraday, gave the Christmas lectures every year for three and a half decades, culminating with a series during the holidays of 1860–61 titled "The Chemical History of the Candle." In introducing his lectures, Faraday told his young audience that "there is no better, there is no more open door by which you can enter into the study of natural philosophy [science]."

This Christmas series marks a milestone in the popularization of science. Japanese translations of the lectures have been read and given as presents ever since. They have gone through more than seventy editions and are still recommended reading for Japanese schoolchildren.

Faraday's vision was based on the classical physics of the day. At that time even the idea of atoms and molecules was not accepted by all scientists. Today this particulate picture changes our view of what goes on when a candle is lit.

When you hold a flame near the wick of a Christmas candle, its heat raises the solid wax's temperature. Viewed at the microscopic level, this increases molecular restlessness and agitation. Once the molecules in the wax have sufficient energy to break loose and tumble over each other, the wax melts. We can detect this process from the change in the appearance of the wax from white to transparent. In the former state, the wax molecules are organized into myriad regions that scatter light; in the latter, they are disorganized.

Peter Atkins, a chemist at Oxford University, points out that Faraday actually sidestepped some fundamental questions posed by the candle flame: Why does it have to be ignited with a match? What drives the chemical reaction?

The driving force is actually the second law of thermodynamics, which can be roughly translated as the tendency of things to become disorganized. In the case of the Christmas candle, it represents the tendency for matter and energy to be dispersed and dissipated while burning in the form of carbon dioxide, water, vapor, and heat. The match is required because, despite this natural tendency for the chemical reactions of burning to occur, there is an energy barrier between reactants in the candle and the products of combustion. Think of this barrier as something like an initial investment that must be made for a candle to enter the energy business. The combustion reaction will take place only if the reactants can be given help vaulting this barrier — by importing some energy with a match to snap chemical bonds and clash chemicals together so they react.

The heat of the flame raises the molten wax's temperature so that its molecules have enough energy to escape into the air. Wax molecules, like all hydrocarbons, consist of a long chain of carbon atoms (between twenty and thirty-five in this case) fringed with hydrogen atoms. In the vapor state, these molecules absorb even more energy when they come into contact with heat from the lit match and vibrate so violently that they shake apart, splintering into smaller molecules — a process called pyrolysis. The fragments react with oxygen in the air and produce fire.

Fuel vapor and oxygen mix at high temperatures at the surface of the flame, releasing heat as new chemical bonds are forged between oxygen and the carbon compounds of the vaporized wax. The resulting rise in temperature melts and vaporizes even more wax to nourish the dark heart of the flame. Simultaneously, oxygen from the air moves toward the flame's surface by diffusion and convection. An uneasy balance between these two forces shapes the flame's surface.

The process of burning is complicated by the generation of loose atoms of hydrogen (H), oxygen (O), and their combination, hydroxyl (OH). These chemical species called radicals, as their name suggests, have the ability to cause chemical mayhem.

In their usual chemical state, as part of a molecule, these entities share two electrons with another species. Radicals, however, have a single unpaired electron that is itching to team up with another electron in a chemical bond. Some of the chemical reactions of the radicals satisfy the electron's urge to be paired, while others create more radicals, which hastens the breakdown of fuel. For example, freed hydrogen atoms bind to oxygen from the air in a reaction that releases great heat. The overall result is water vapor and carbon dioxide gas. Both rise from the flame, because hot, less dense combustion products ascend by the process of convection. This imports oxygen and exports combustion products, drawing the flame out into the characteristic teardrop shape.

Examine the flame of an Advent candle, and you can see that just above and around the burning wick is a dark cone topped by a luminous yellow region. The temperature in the dark cone near the wick is fairly low, 600°C, and rises to about 1,200°C at the heart of the yellow region. The highest temperature, 1,400°C, is found on the edge of the yellow portion of the flame.

The vaporized wax hydrocarbon molecules are decomposed by the heat in the dark cone near the wick. Around the yellow zone of the flame, radicals from the decomposed hydrocarbons react with oxygen from the air to form carbon dioxide and water in a complex, not fully understood way.

A spoon can confirm these processes. Hold it above the flame and notice how water condenses on the cooler metal. Place it in the yellow zone, and you can sample soot. Above the wick in the dark zone, vaporized hydrocarbon will condense to form a thin layer of wax. This was one of the many elegant demonstrations used by Faraday in his Christmas lectures.

Candlelight

Given the ancient preoccupation with flames in the cold of winter, it is not surprising that most of the radiation of a burning candle is released as heat, or infrared light. Faraday was at his most poetic when discussing the outstanding trace of visible radiation: "You have the glittering beauty of gold and silver, and the still higher lustre of jewels, like the ruby and diamond; but none of these rival the brilliancy and beauty of flame."

The luminous yellow zone of the flame is responsible for most of the light produced by a candle. This is also called the carbon zone. The soot particles that lurk there are composed of carbon atoms that have been released from their original chains in the wax to link themselves in arrays of various sizes, from ten to two hundred nanometers, and then into chains. The soot region also contains one of the most fashionable molecules of the moment — a form of carbon discovered only recently, called buckminsterfullerene, where sixty carbon atoms link up to form what looks like a football.

How can these carbon footballs and other components of black soot become so incandescent? Soot particles do indeed glow like any other hot matter, but "you would hardly think that all those substances which fly about London, in the form of soots and blacks, are the very beauty and life of the flame," Faraday remarked.

The physics of his day could not account for the yellow and orange colors of a Christmas candle. This would require an understanding of events at the atomic and molecular level. It would take another half century for primitive models of the atom to emerge. Among these was the "plum pudding" model proposed by J. J. Thomson in 1904. His model incorrectly depicted an atom as consisting of negative particles embedded like plums within a sphere of positive electricity of uniform density — "a number of negatively charged corpuscles in a sphere of positive electrification."

Today, thanks to the quantum theory pioneered mainly by European physicists between 1900 and 1928, the atom is understood to be a tiny positive nucleus surrounded by a mist of negatively charged electrons. The shape and extent of the mist reflects the energy levels of the electrons. When heated, carbon atoms in an Advent candle's soot move about, twisting, jumping, and distorting when electrons hop around between energy levels. Some of this energy converts itself into particles of light which fly out of the flame, varying in color (frequency) according to the size of the energy hops. The classical physics of Faraday's day worked reasonably well for low frequencies, at the red end of the spectrum. But at high frequencies, or the blue end, it predicted that a heated body would send out an infinite amount of energy. This absurd result was dubbed the "ultraviolet catastrophe."

The correct explanation would not come until the turn of the century, when Max Planck at the University of Berlin launched the field of quantum physics. To prevent this "catastrophe" and explain how hot bodies such as Advent candles radiate light, Planck had to assume that the atoms in the soot could vibrate only in multiples of a basic quantity, found by multiplying a tiny constant by the frequency of the light. Then in 1905 Albert Einstein realized that light could exist only in little chunks, called quanta, a view at odds with the prevailing picture of light as a wave. Two decades later the American chemist G. N. Lewis christened the quantum of light the photon. The more energy in the light quanta, the bluer the light. For example, the pale blue zone at the bottom of a flame is due to quanta sent out by "excited" versions of water, carbon dioxide, and other molecular fragments which are agitated because their component electrons have been promoted to a higher rung on a ladder of molecular energy levels. When these electrons drop down to a more stable level, the discarded energy is sent out in the form of photons.

The Mystery

In one respect the candle seems almost as mysterious now as it did in the days of Faraday's splendid Christmas lectures — if you are an astronaut, that is. For years scientists have puzzled over the behavior of candles in the absence of gravity to highlight the influence of the planet's gravitational tug upon a living flame. Would a cosmic Advent candle burn indefinitely in a large volume of quiescent air or be snuffed out by a surplus of combustion products and oxygen starvation? At the heart of the question is the way in which the absence of gravity affects the flame's supply of fuel and oxygen. On Earth molecules diffuse and are moved by convection, as the hotter air rises and cooler descends. In space, however, the process of convection is switched off. The flame should burn, but only as long as the oxygen molecules diffuse in quickly enough and the hot fuel diffuses out.

Scientists are divided as to whether the flame of a Christmas candle could survive for very long in zero gravity. Some argue that it could, because the equations governing the process of diffusion say so. Others believe that diffusion is too slow a process to nourish the flame. Experiments on Earth, by literally dropping the candle to switch off the effects of gravity, have proved inconclusive. Another method to switch off gravity was adapted by Howard Ross and Dan Dietrich at NASA's Lewis Research Center in Cleveland, working with James T'ien of Case Western Reserve University. They studied a candle flame aboard a shuttle mission that flew in 1992. Immediately after being lit, the candle flame was spherical and had a bright yellow core. After eight to ten seconds, the yellow (presumably from soot) disappeared, and the flame became blue and hemispheric, measuring only half an inch or so across.

In the absence of gravity to define an "up" or a "down," a space flame tends toward sphericity. Heat loss to the wick of the candle, however, quenches the base of the flame, causing it to be hemispheric. In microgravity the process of diffusion controls the supply of oxygen and fuel vapor to the basin-shaped flame. Being a slower process than convection, this significantly reduces heat generation, so the flame temperature is lowered to the point that little or no soot forms, as evidenced by the all-blue flame on the shuttle.

Further investigations that confirmed this picture were carried out aboard the Russian space station *Mir* in 1996 by American astronaut Shannon Lucid. So far the studies seem to suggest that a candle can burn in space. But the quest to understand the Christmas candle is far from over, for there remain many aspects of the zero-gravity candle flame that puzzle researchers. One is the discovery of strange flickerings just before a space candle is extinguished.

However well we come to understand its mysteries, the flame retains its inspirational quality. This quality was expressed by Faraday, who, echoing the prayers familiar from Christmas services, hoped that we might all be compared to a candle: "That you may, like it, shine as lights to those about you; that, in all your actions, you may justify the beauty of the taper by making your deeds honourable and effectual in the discharge of your duty to your fellow men."

4

GIVING AND SHOPPING

The only gift is a portion of thyself. Thou must bleed for me. Therefore the poet brings his poem; the shepherd, his lamb; the farmer, corn; the miner, a gem; the sailor, coral and shells; the painter, his picture; the girl, a handkerchief of her own sewing. . . . It is a cold, lifeless business when you go to the shops to buy me something, which does not represent your life and talent.

RALPH WALDO EMERSON, ESSAYS, 1844

EVEN THE APPARENTLY INNOCENT ACT OF GIVING a Christmas card groans with meaning. According to psychologists, there is a hidden language in this seasonal ritual that reveals a great deal about our nature and our relationship with others. The same is true, they say, for giving presents and the way we wrap them. Under that tree lies tantalizing evidence of how we relate to friends and family.

Meanwhile, anthropologists tackle the paradox of the seasonal spending spree. We moan endlessly about the "materialism" of the holiday, the consumer hype, the hard-sell advertising, and how Christmas displays seem to appear in stores earlier each year. Every season the media reiterate what has become the cliché of all Christmas clichés: the growing commercialization of the holiday threatens the very spirit of Christmas.

Yet the shopping intensifies each year. The demand for Christmas goods starts in late summer, and, by one estimate, Christmas purchases now account for one-sixth of all retail sales in the United States. One study concluded that the holiday ritual is "the most sociologically significant gift-giving in modern American culture."

In Britain around 8 percent of the economy is devoted to producing items that will be given away, with Christmas gifts accounting for 4 percent of an individual's annual income. In Japan, where shoppers scurry around during the *oseibo* gift-giving season to the refrain of *Meri Kurisumasu,* the proportion is probably greater.

How can Christmas be, at one and the same time, a great religious festival and the biggest commercial holiday in the Christian world? This is no accident, argue anthropologists. Shopping and giving are part and parcel of the Christmas festivities because they help to reaffirm family values in an increasingly materialistic and impersonal world.

Some parties, most notably the American organization SCROOGE (Society to Curtail Ridiculous, Outrageous, and Ostentatious Gift Exchanges), have lobbied for spending on self-improvement lessons, smoke alarms, first aid kits, and other "sensible" gifts. But as one academic points out, this cold-eyed, rational view of Christmas could do more to demolish the romance of the event than the annual ritual of mass consumption itself.

A Brief History of Giving

The language of giving is ancient. In the animal kingdom, gifts are given for good reason. In many insect and some bird species, females accept males as mates only if they are offered a morsel of food. Collecting such a gift may be as hazardous for males as the hunt for a female itself. For example, male scorpion flies mate only if they offer females nuptial gifts of palatable insect prey. In searching for these presents, the males run the risk of becoming entangled in the webs of predatory spiders.

Their generosity may seem baffling when viewed in the light of Darwin's theory of natural selection, which explained, at a single blow, the evolution of all sorts of adaptations that enhance the animal's ability to survive against the backdrop of a nature red in tooth and claw. Darwin was himself puzzled by behaviors that do nothing to increase survival and that may even hamper it. He concluded that features such as the peacock's tail — characteristics usually sported only by males — evolve by a process called sexual selection. Individuals possessing these characteristics, which are desirable to the opposite sex, acquire more mates than their competitors and thus have better chances of reproducing.

Robert Trivers of the University of California at Santa Cruz went on to argue that although both sexes attempt to pass as many copies of their genes to the next generation as possible, females and males have evolved different strategies to achieve this same end. For those "generous" birds and insects, gifts therefore have a well-defined purpose: to boost the chances of reproduction. Human beings also give for a purpose and have been doing so since the emergence of hunting technology some fifty thousand years ago. Darts and other projectile weapons enabled hunters to tackle game that was so big — a mammoth, for instance — that sharing the spoils became mandatory.

Gifts are never merely gifts, however. Only part of the reason for our generosity is to be nice to our fellow man and woman. We expect something in return.

Giving at Christmas is charged with a particular meaning. In 1252 Henry III entertained a thousand knights and peers at York. The Christmas feast was so costly that the local archbishop donated 600 oxen and a vast sum of money (£2,700) for the feast. Nearly a century and a half later, Richard II provided 2,000 oxen and 200 tuns of wine for the 10,000 people who dined daily at his expense. Edward IV fed more than 2,000 people each day over the Christmastide of 1482 at Eltham, southeast London. This lavish hospitality was extended for very good reason: to forge alliances and strengthen community bonds. In this way the kings underwrote feudal relationships. Part of the seasonal package was for the powerful to attend to the needs of the powerless, an inversion that served to ease tensions between the strong and the weak and thereby reinforce the status quo.

Gift giving in preindustrialized societies was seen by the French ethnographer Marcel Mauss as a way of establishing social contracts with strangers. In his *Essai sur le don* (1925), regarded by anthropologists as a milestone work, he outlined how this practice served a useful purpose in the days when there was no state police or militia to keep the peace. Mauss reminds us that there is no gift without bond, without bind, without obligation or ligature.

The science writer Matt Ridley brought the idea of gift exchange and reciprocity to life in his description of a people called the Hazda, who live in Tanzania. In the wooded savanna near Lake Eyasi, they forage for honey, roots, and berries and hunt large animals such as giraffes. When it comes to big game, they either return empty-handed, which is usually the case, or they down an antelope or giraffe with bow and arrow, only to give most of the meat away. What they get in return for their generosity is kudos and the kind of female admiration that could pave the way for extramarital affairs, underlining the fact that anthropologists don't quite equate tit for tat with reciprocity: the former means swapping similar favors at different times, the latter sharing different favors at the same time.

The Hazda hunters are converting giraffe meat into a durable and valuable commodity — prestige — that will be cashed in for a different currency of advantage at a later stage. Ridley sees in these actions distant echoes of the origins of modern markets in financial derivatives: "He [the hunter] is entering into a contract to swap the variable return rate on his hunting effort for a more nearly fixed return rate achieved by his whole group."

This kind of thinking transforms the meaning of the gift from one of social sentimentality to one of impersonal economics that follows culturally prescribed laws and obligations. Indeed, in some societies the sense of obligation attached to the reciprocal gift was so strong that the gift could be used as a weapon.

Until the nineteenth century in the Pacific Northwest, "first nation" people would attempt to belittle their rivals by the ritual of the potlatch. In a society obsessed by status, competitors staged a bizarre battle of generosity in which blankets, oils, food, pelts, canoes, and decorated sheets of copper were given away. Every present meant another rung up the social ladder, and every failure to reciprocate cost status in these escalating gift wars. Many commentators have remarked on how this competitive gift giving sounds rather like the destruction of wealth that goes on every Christmas in our own society. Three decades ago Marshall Sahlins noted that the closer the kinship between the giver and the recipient of the gift, the less emphasis was placed on reciprocity and the more on sentiment. Within the family, there was little tally kept on who spent what on whom. Within the larger social grouping of the tribe, however, a certain amount of diligence was required to ensure that you gave as good as you got. This same reciprocity ruled when it came to gifts between unrelated allies. But in the case of rival tribes, the idea was to try to get more out of them than they got out of you.

Echoes of what Sahlins recorded were observed in Middletown, a midwestern community adored by ethnologists. There Christmas gift giving was studied by Theodore Caplow of the University of Virginia. He found that the vast majority of presents went to those within the nuclear family. Gifts to relatives outside the core family were less common and more conditional. For collaterals — siblings, siblings' spouses, siblings' children, and parents' siblings — giving was even less common.

Gifts within the core family were given without the expectation of an equivalent return, but Caplow found a growing emphasis on reciprocation when it came to collaterals and their spouses. Beyond the circle of kin, whether teacher or garbage collector, mostly small gifts, which amounted to little more than tips, were given.

Another study by Caplow, based on interviews with 110 people in Muncie, Indiana, found that when people did not give to someone who had previously been on their list of recipients, it meant that they expected the relationship to wane in importance, even end altogether. Thus the nonappearance of a gift can signify the end of a friendship.

The Psychology of the Gift

The previous section underlines how the giving of gifts is a dialogue of intimacy, a language for the communication of emotions and values. We give gifts to promote the reputation that we are nice, generous people. And we do it to put pressure on the recipient to reciprocate. When, for example, you take a basket of fruit to a friend in the hospital, it is because you would like him or her to do the same for you.

Gifts have the power to make or break a relationship, for they are indices of how we interpret the status, power, taste, and emotion of our peers. According to Adrian Furnham, a psychologist at University College, London, they reveal how socially aware we are in perceiving others. It's not just the issue of whom we choose to give presents to or how much or how little we spend on those presents, but what sort of gifts we select. And when our motives for giving a particular gift are incorrectly interpreted, our faux pas is on display for all to see. For example, the gift of a fluorescent fluffy toy might be thought an insult by someone who perceives himself or herself as sophisticated.

"As a channel of communication [a gift] has limited capacity because the range of messages is few and the language not well known," Furnham says. "Perhaps the gift-phobics who discourage the exchange of gifts between family and friends do so because they don't speak the language and agree with Wittgenstein, who so wisely noted: 'Whereof one cannot speak, thereof one must be silent.'"

Psychologists have now started to decode the language of gifts in an effort to unwrap this seasonal ritual. They have studied different gift occasions and assessed the various stages in the gift-giving process, the function of gifts, and the norms that govern who may give what to whom and why certain gifts — for instance, money — are often considered inappropriate.

Men and women behave very differently when it comes to Christmas gift giving. At the University of Winnipeg in Manitoba, David Cheal had great difficulty interviewing as many men as women for one study of Christmas gift giving. The reason soon became clear: women remain the principal actors in gift transactions. The annual hunt for that ideal present is overwhelmingly seen as women's work. Indeed, among couples it is usually the women who maintain the gift economy.

Men tend to give more valuable gifts, less often. Part of the reason is that men generally earn more than women. But women have been said to dominate Christmas giving perhaps because it is seen as a family festival and women are the "kin keepers," taking more responsibility for maintaining family and social ties. One of Cheal's respondents explained that her reason for giving is "to be a message. You have interest in that person. You have concern. You have love, affection, whatever the message is at the moment."

Other studies have shown that we are little different from the Hazda in that gift giving often puts the recipient under an obligation, exploiting a reciprocal instinct that places the act closer to pure barter. Much of Christmas giving actually reveals a more calculated character, following certain rules and obeying certain taboos. "To violate these rules, to give too little, or indeed to give too much, can be insulting," say Carole Burgoyne and Stephen Lea of the University of Exeter, England.

One traditional taboo is the gift of money. If we hand out checks and cash, the materialist underbelly of Christmas is laid bare for all to see. As a result, money is not a universally acceptable medium of exchange.

Gifts of money also imply a lack of effort and insight on the part of the giver, according to a study of 92 students conducted by Burgoyne and David Routh of Bristol University in England. Another study by Lea showed that this was particularly so when money was given by a child to a parent, but not when it was a gift from a grandparent or parent to a child.

Today's psychologists see gifts as a way of initiating and maintaining relationships — just as we observed in the case of Henry III. According to Burgoyne, Christmas tends to differ from other giving rituals, such as birthdays, because it is more likely to involve a simultaneous exchange. In relationships where reciprocity is expected, there can be serious consequences of the failure to give a gift. These are more likely to occur in closer relationships, such as between siblings, parents and children, or girlfriend and boyfriend. "The nonappearance of a gift is likely to lead to broken relationships and family rows unless there is a very good explanation for it," Burgoyne says.

Nevertheless Christmas is an occasion for the relaxation of other rules of gift giving. Because seasonal presents are handed out more widely, they are often less intimate and personal than birthday presents. This, of course, can be an advantage for those who are trying to start up a relationship. But gently does it, warns Burgoyne: "Gifts that are too expensive may signal a level of commitment and impose a sense of obligation that is not wanted by the recipient. Thus an inappropriate gift — one that is either too cheap or expensive — or one that seems to expose a lack of taste on the part of the donor — carries the risk of rejection."

Gifts are also an excellent way of atoning for sins, but they may be rejected if judged as not sufficiently compensatory. If they are *too* compensatory, however, they also can cause an offense. Precise reciprocity could be seen as an unfriendly act if one breaks the rule of approximate worth — that is, giving a return gift of approximately the same monetary value.

The Psychology of the Christmas Card

The most common gift of all is the Christmas card. Henry Cole's pioneering card paved the way for a vast variety, some dressed in satin and silk, others gilded and frosted, and still others fashioned in the form of stars, crescents, and fans. In the United States, the first card appeared sometime between 1850 and 1852 in the form of an advertisement for "Pease's Great Varety [*sic*] Store in the Temple of Fancy." The card was adorned with Santa and a black slave laying the Christmas dinner table. By the 1880s manufacturers began producing cards cheaply, enabling anyone to afford them. One example was a set of cards called the "Penny Basket," introduced by the German company S. Hildesheimer & Co., which had branches in London, Manchester, and New York. By 1883 the *Times* in London noted that this "wholesome custom" had become a way to end strife, mend relationships, and strengthen family ties.

Today's psychologists would agree. Cary Cooper, an American professor of psychology working at the University of Manchester Institute of Science and Technology, in England, argues that the act of sending a card is laden with meaning, particularly now that our ever-growing commitment to work makes the Christmas break more significant than in previous years.

Three decades ago the ritual of exchanging cards was little more than a minor social obligation. Today it has become much more. After the "I-driven" culture of the 1980s, "we are entering an era where people are thinking more about other people," Cooper says. When people send cards, they are reaching out to those with whom they have lost contact or want to sustain a relationship. "It is a form of social networking," he notes.

In a time when many people no longer live near their extended families, don't know their neighbors, and are highly mobile, cards provide social glue. "The cards I get from the United States don't just have signatures but contain letters and pictures," Cooper says. "It is a way of saying, 'Hey, I haven't forgotten you. I still value your friendship.'"

Just like the decisions we make about whom to vote for, whom to marry, and what to wear — things we think of as purely rational, individual matters — the Christmas cards we send and the way in which we send them may, in fact, say a lot more about us than we would care to admit. Ironically, this factor may supply another reason the custom has become such a central feature of Christmas.

In a recent analysis, Adrian Furnham and Ruth Leigh of University College London point out that there are many things going through our minds when we devise a list of whom to send cards to. A few people keep the previous year's cards so that they can repay their debts. Some people scour their address books for all those with whom they have fallen out of touch. Others fit the chore in among the obligations to shop or decorate, or simply reply tit-for-tat as cards arrive in the mail.

Any deviations from a straightforward exchange reveal a lot about status, argue Furnham and Leigh. People to whom you sent cards yet received none in return are likely to be above you in the social ladder — those whom you want to cultivate as friends. Those from whom you received cards but did not reciprocate are probably below you in the great social pecking order. They are "likely to be attempting to ingratiate themselves with or gain favors from you yourself. Of course this is more exaggerated in individuals who are upwardly socially mobile."

What is written on the card also is revealing. A long, windy message is often penned by someone trying to impress. A little joke may attempt to make up for the long delay in reestablishing contact. A simple signature is saying the sender is much too busy and important to bother with pleasantries.

The type of card chosen betrays a great deal about who sent it.

- A homemade card may say, "I am wealthy and have enough leisure time to make my own cards." Or it may declare, "Admire my artistic bent." Or it may coo, "I am spending all my time on your card to show you just how much I care for you."
- A "green" card — say, featuring a large-eyed seal, winsome elephant, or leaping whale — printed on recycled paper

and carrying a greeting in a foreign language, can not only establish the sender's environmental credentials but also suggest that the recipient should feel a twinge of guilt for sending that commercial card on nonrecycled paper adorned with reindeer, Santa, or a mouse peeking out of a stocking.

- A privately printed card reeks of money. But it also may suggest that the sender is too busy to fiddle around with signing a lot of cards. An interesting variant is the photograph card, which is calculated to show off that new house or addition to the family, whether car or baby.
- The institutional card may take various forms. Firms housed in particularly ugly buildings may have to resort to a generous amount of snow to hide the fact. Others may choose to adorn their cards with some element to announce the business they are in, whether microchip, widget, or spiral of DNA. "This has much of the same function as people who wear allegiance ties, cuff links, blazer badges, or some other form of insignia, stating publicly their identification with the values and aims of an institution," Furnham and Leigh explain.
- The political card is used as a badge of social belief, mostly exchanged between small groups of stalwarts.
- The cartoon card is for those who have no desire to remember what Christmas is all about.
- The commercial card, with its Dickensian scenes, mythical figures, Europeanized biblical scenes, and anthropomorphized animals, offers the advantage of relatively cheap diversity — to the point of even carrying pagan inscriptions. It is "a golden opportunity for the rugged individualist who wishes to escape from snow scenes and Victorian gentlemen in top hats," Furnham and Leigh note.

- As for the microchip musical card, "their tinny warblings can only be stopped by violent means, and to make things worse, the picture will be of either a fluffy animal or a badly drawn snow scene. Sever all relationships with people who send these cards," the psychologists suggest. "There is no hope for them."
- And, of course, the spirit and influence of Dickens are alive and well in the form of the charity card. This genre, akin to the "green" card, advertises the generosity of the sender — even, heaven forbid, that they are suffused with the true spirit of Christmas, brotherhood, peace, and generosity. Because such cards enable one to do so many things simultaneously, they are becoming the most popular type.

Finally, the way you display your cards is revealing. The fact that they are on show at all is a way of announcing to visitors how many friends and acquaintances you have. A century ago, the calling card of any rich or famous person would be left in the hall for other visitors to see. Today the same favor is bestowed on celebrity cards, which are prominently displayed on the mantel alongside those from anyone who may happen to pop in to offer Christmas greetings in person.

An Anthropological Christmas

Anthropologists believe that *they* are the ones who really understand the orgy of giving. In a milestone study, *Unwrapping Christmas,* edited by University College London professor Daniel Miller, some anthropologists point out that the emphasis Europeans and Americans now place on emotionally charged gift giving emerged in the mid- to late 1800s. This change coincided with the period of intensive industrialization and a growing sense of the domestic sphere as a special moral realm distinct from the harsh realities of the workplace. This emphasis on "the family Christmas" emerged, spurred on by the enormous popularity of Dickens's *A Christmas Carol,* for example.

Buying manufactured goods as Christmas gifts did not become widespread in the United States until the 1870s. Leading department stores such as Macy's in New York began to lure customers with striking displays of imported dolls and other exotic products. Santa Claus also began to appear in department stores, cathedrals of the culture of consumption after the Civil War. "If Santa is the god of materialism, what better place to enthrone him than in the department store, which did so much to foster consumer culture," comments Russell Belk of the School of Business at the University of Utah.

We are now living in a commodity culture, one in which we use money to buy impersonal, mass-produced objects made for profit by people we've never met. As James Carrier of the University of Virginia puts it, "These objects are unsuited to what people want them to mean and to be, for they bear no human identity or being. They are manufactured in the world of work and express only the impersonal desire for profit by the company that made them and the impersonal acting out of work roles, lightly adopted and in return for money, by the unknown people who produced them."

The challenge we face at Christmas, argue anthropologists, is to transform the mass-produced objects found in stores into gifts by scrubbing them clean of the contamination of money.

There are various ways that people build meaningful social relationships in a world where we give friends and family anonymous objects. The first is to mutter, "It's the thought that counts," and place more emphasis on gifts that are frivolous rather than useful. It is better to purchase luxury items than all those "necessities" bought at other times of the year, the argument goes.

Another way to add meaning is to wrap the gift in beautiful paper and tie it up with ribbons and bows. This strategy temporarily makes the object irrelevant and the act of giving the focus of attention. Claude Lévi-Strauss argues that wrapping overlays the mass-produced product with sentiment and the giver's identity.

But perhaps the most shocking implication of commodity culture is that we had to invent the ritual hell of Christmas shopping as part of this cleansing process. James Carrier believes that shopping is an integral part of the experience rather than "an unfortunate commercial accretion on a real ritual and familial core." In this respect, it is like the ritual of family cooking, which converts store-bought groceries into a meal that expresses, embodies, and strengthens the family bond.

The next time you whine about the crowds of shoppers in December, just remember that the seasonal spending spree is meant to be unpleasant, hard work. Like the British and Germans, "Americans commonly see Christmas shopping as an onerous task," Carrier says. "People regale each other with stories about how hard it is and resolve to start earlier next year." Christmas has to be worked at if it is to be done properly.

Shopping is a pivotal part of Christmas. If people endlessly grumble about what hard work the shopping is and about growing commercialization, it is because, Carrier notes, these complaints help to affirm that at least once a year, it is possible to "wrest family values from recalcitrant raw materials." Carrier goes on to say that this is why children give few gifts: "they do not normally undertake the daily round of shopping, transforming, and giving objects."

Theodore Caplow's Middletown study in America revealed that few gifts were handmade, and most of the handmade items were presents from young children. In Sweden homemade gifts, red drapes, and home-pickled cowberry sprigs are de rigueur, precisely because people are most concerned about "folk culture" and a sense of national identity and origin. By contrast, in America and Britain, homemade gifts are eschewed, because there the concern centers on how to make "family values" compatible with rampant commercialism.

In the end, all the consumer hype and advertising that accompanies the seasonal spending frenzy "allows us to show ourselves how important consumption is," says Marilyn Strathern, professor of social anthropology at the University of Cambridge. "We glut, we overdo it, we grumble about how commercial it all is, but we are reflecting back on the value of it all."

Big Brother Is Watching You Shop

Anthropologists are not the only ones who watch your every move as you hunt for that ideal gift. Psychologists armed with computers, scanners, and video cameras have joined them so that stores can extract even more money from you during the Christmas season. A wide range of monitoring equipment records customers' behavior, while retailers experiment with smells, music, and "atmospherics" to encourage people to buy just one more gift.

The most obvious incentive to encourage shopping is to offer lower prices in a Christmas sale. Robert East of Kingston University in southwest London noted that a 10 percent drop in price can boost sales by 20 to 30 percent. If the goods are put on a special display, purchases can go up as much as 80 percent. Add local advertising, and sales can be pushed to 200 percent, although, he stresses, these figures are averages and vary from product to product.

Retailers use various strategies, policies, and procedures in timing their markdowns of Christmas merchandise, adds Dale Lewison of the University of Akron. "Some retailers start taking small and early markdowns before Thanksgiving, while others wait until after the weekend following Thanksgiving — the biggest shopping weekend of the year. Still other retailers wait longer to mark down merchandise."

The longer the retailer waits, the bigger the markdown needed to generate customer interest, he says. Many retailers have an automatic markdown policy every week or two, so that "timing of markdowns becomes quite an art and science if the retailer wants to maximize both revenues and profit."

Stores can use tie-ins with Christmas events, movies, and other hot happenings. They can add a fancy window display with animated Santas, electronic elves, and a mountain of fake snow. Depending on where the store is based and its customers, the seasonal display can include Hanukkah and Kwanzaa paraphernalia.

To find out what effect displays have on consumers, a team from Nottingham University in England used video equipment to conduct a vast study. "The idea was to see how far away people walking along the pavement are attracted to displays and whether they are tempted to enter," says Roy Bradshaw of Nottingham's Retail Analysis Team.

The team has analyzed photographs to produce more than a million observations of pedestrians in city shopping areas and stores. These studies show that as many as five times more people enter some stores than cash register receipts alone would suggest. "Shopping is very often a social activity in which social dynamics play a very critical role," Bradshaw says.

That is why anonymous shoppers were analyzed not only by age and sex but also by the "social association variable." The same person will have different shopping behavior depending on who accompanies him or her. A thirty-year-old mother will behave differently if she is shopping for essential items during her lunch break from work than on the weekend with her partner and/or children. In the first case, it is something of a "commando raid" — straight in to purchase the items on her list and straight back out. In the second case, it is partly a reconnaissance mission, strolling around and seeing what is being offered. When time is not such an important constraint, shopping becomes much more of a recreational activity.

Bradshaw's data, reproduced on computer, show how well stores work — hour by hour, day by day — for the benefit of both customers and owners. They reveal where shoppers look, where they go, the bottlenecks they encounter, and the "sales per thousand customers passing the product." The studies have shown that in some cases as much as a 20 percent improvement in increased sales could be made each Christmas just by altering the layout of the store.

Another "spy" on the Christmas consumer comes in the form of the bar code. This has been monitoring shoppers since 1967, when the first retail system was introduced into a Kroger supermarket in Cincinnati. Six years later the industry agreed on the Universal Product Code, which today is to shopping what DNA is to biology.

The pattern of lines and bars in the code can reveal to a laser scanner the identity of a product: the manufacturer and what the item is. This can be compared with a database to pull out the price and check how many are in stock, for example, says Sunil Gupta, who conducts bar code research at Columbia University.

Using a computer, bar code information can be combined with knowledge of the layout of the store to highlight hot spots of activity. Not surprisingly, such analysis shows that eye-level height is the most profitable display area for a product, giving about twice the return of placement at a lower level. In addition, demand can be linked to weather, so that it is possible to show to what degree a cold spell can influence the buying of warm clothing. Gupta has shown how predictable seasonal sales hurt brands in the long run, as customers then refuse to buy them at the regular price. Eventually sophisticated computer models will be used by major stores to predict what items to stock and promote each Christmas by drawing on a vast database of information ranging from weather to customer demographics. Gupta says that department stores are only just beginning to recognize the power of bar code data.

Big Shrink Is Manipulating You

The ways stores are designed and gift ideas are presented are based on a thorough understanding of consumer psychology. This makes the average shopper seem like a hapless pawn who is ruthlessly manipulated by multimillion-dollar businesses. But, says Mark Uncles of the University of New South Wales in Australia, "this is not always as deeply sinister as it might appear."

Although it is true that designers and retailers attempt to influence the moods and emotions of shoppers, sound operational considerations and customer needs also must be taken into account. For instance, wider aisles are needed because the size of shopping carts has increased, and that has happened because more people shop less frequently and in larger quantities when they do shop. It isn't simply that wider aisles encourage customers to browse and buy more at Christmas. Similarly, it is more convenient to have in-store bakeries and fish departments against the back wall of a store. It isn't just that these services draw shoppers through the store and stimulate impulse buying.

Nonetheless, retailers do explore how they can help shoppers part with more of their money by manipulating store atmospherics and design, from colors, smells, and lighting to layout, the positioning of displays, posters, in-store "theater," and music.

Cosmetics and jewelry stacked near a department store's entrance create a pleasant smell and "put a sparkle in people's eyes." Supermarkets display fruits and vegetables at the front of the store for similar reasons. Supermarket layout often steers shoppers clockwise: "psychologists say we have a predisposition to turn right on entering closed spaces," Uncles explains.

The use of the smell of coffee and freshly baked bread to lure customers goes back at least as far as the eighteenth century, while the use of fresh herbs is even older, possibly dating from the Middle Ages, if not before. Today many large stores on both sides of the Atlantic pump the smell of fresh bread from the ovens to the entrance of the store to attract customers and set stomachs rumbling.

In the weeks running up to Christmas, the scent of cloves, cinnamon, brandy, and other "jingle smells" may waft through the aisles. One Christmas Woolworth stores in Britain released each quarter of an hour a puff of mulled wine dissolved in carbon dioxide from a small canister. A panel of staff had decided that mulled wine created the right ambience to put customers in a suitably mellow and festive mood. The Tate Gallery in London, meanwhile, plumped for the smell of brandy.

Although it has been used for hundreds of years, smell has been remarketed as the "hidden persuader" in the light of research that suggests the right smells might even encourage consumers to spend more money. One study of sales at a women's wear retailer in the United States found that there was a 15 to 20 percent increase in the average amount customers spent when the store was filled with a peach aroma, compared with the previous year when no scent was used. Others have seen earnings in a Las Vegas casino jump by 50 percent when a pleasant smell was wafted over gamblers, and shoppers were found to be more willing to buy Nike sneakers, and even to pay more for them, in a floral-scented room. Indeed, smells might have other effects: one company in Britain manufactures synthetic human pheromones, or signaling chemicals, that are believed to make people feel more relaxed.

Smell is not the only sense that is being manipulated by retailers. Colors are likewise thought to help people part with their money. One study revealed that people moved closer to red, which is more arousing than "cool" colors such as blue and green. That led to the suggestion that red is appropriate where quick decisions will benefit the store. Could that be why it is Santa's favorite color?

To focus attention on Christmas products, stores play seasonal music, especially carols. The pace of the music can be important. Low-tempo "easy listening" music relieves stress and encourages shoppers to take more time. One study found that slower music boosted sales by 38 percent compared with faster music, presumably because shoppers lingered longer. That would suggest stores should make more money playing "Silent Night" than "Jingle Bells."

The type of purchase can be influenced by the musical genre, according to a study conducted in a supermarket by Adrian North, David Hargreaves, and Jennifer McKendrick from the Music Research Group at the University of Leicester, England. French wine flew from the shelves on days when French accordion music played on the in-store sound system: five times as much French as German wine was sold. But German wine triumphed on days when thigh-slapping Bierkeller oompah music roared forth, with twice as much German as French wine being sold.

The study supports what the psychologists called the "preference for prototypes" model, meaning that "if you hear music that you label as French, you start to think of things about France in general," North explains. Other studies have underlined the influential role of music. Classical music prompted the purchase of more upmarket wines than pop music, for example. When it comes to departments where younger people tend to shop, the volume should be turned up. But there has to be a limit. The Leicester team is now studying the long-term effects of Christmas carols, notably the irritation caused when shoppers are subjected to them again and again.

Atmospherics also can be used to "demarket" a product, even at a time like Christmas. The classic example is the state-run liquor store, which is deliberately made uninviting in some countries to discourage drinking. You might think that Christmas crowds would have a similar effect, creating logjams of customers, fighting in the aisles, or even "shopping cart rage." Robert East points out that the impact of store congestion is subtle.

Crowds are undoubtedly worst during Christmas shopping. But that is not necessarily a bad thing. Whether we like crowds or not depends on where we are: yes if it is a bar, no if it is a bank, says East. When it comes to a store, too few people may advertise the fact that it has poor products or high prices. Crowds in stores can make us more purposeful and efficient in our shopping. Equally they could make stress hormones soar to unbearable levels.

The Laws of Christmas Shopping

An ambitious project to tease out the details of how we hunt for that special Christmas present has been under way at Concordia University's Department of Marketing in Montreal. There Michel Laroche, Chankon Kim, Gad Saad, and Elizabeth Browne put forward eight propositions that govern whom we consult about buying a gift and how long we spend searching for it.

The group was interested in studying the effects of a range of factors: "experiential pleasure," which is the sheer enjoyment of browsing during the gift quest; the shopper's "social risk reduction strategy" — for instance, how much effort we invest in selecting a gift to avoid appearing cheap or insensitive; and "psychological risk," the strength of our resolve to cope with the misery of fighting with a mob of shoppers.

Drawing on earlier research and a great deal of common sense, they proposed the following prototype "laws of Christmas shopping," focusing in particular on the influence of information — from browsing, examining potential gifts, questioning clerks, and so on — that a store makes available.

1. The shopper will spend less time hunting around if he or she is accompanied, but more time when searching for a costly gift or when adhering to a tight budget.
2. The less time we have to shop, the more we rely on clerks, information desks, and shop assistants, and the less we rely on browsing, examining labels/packaging, and so on.
3. The use of predetermined gift selection — orders from a "significant other" for a particular brand of perfume, for example — should "result in less in-store general information search, and more brand- or item-related information search."
4. The closer your relationship with the recipient of a gift, the more likely you are to know what he or she wants and the less time you will spend rooting around in a store.
5. The more difficult the recipient — for instance, a much-disliked mother-in-law — the more time you will spend finding the appropriate gift.

6. Those who love shopping, bargain hunting, and giving will invest more time in the ritual, while those who prefer buying generic items to specific brands will invest less time.
7. Demographic variables such as age, education, and family size also will influence the hunt for presents. Women, for example, will tend to search for more in-store information than men, the latter relying more on the advice of sales-clerks. The researchers found plenty of evidence in the literature to support the common anecdotal observation that male Christmas shoppers are a sorry lot: "Men will tend to leave their Christmas shopping until the last minute, then will enter a store, list in hand, desperate," says the Concordia team. Or, as an earlier study noted, "males were distinguished by . . . their desperation, abrupt-ness, tardiness, and discomfort with the ritual process." (I plead guilty on all counts.)
8. The reliance of shoppers on displays — one example of what the team refers to as a "non-personal in-store information source" — is more influenced by the location of the display in the store, for example, than by demographics such as the social class and potential spending power of the shopper. When it comes to consulting a sales assistant, gender or social status matter less than "situational variables," for instance the time available when seeking an expensive gift, such as a gold watch, among a vast selection of pricey timepieces.

Once the Concordia team had prepared these propositions, they put them to the test. Socks, shirts, scarves, and so on had been identified as a popular gift choice by an earlier study on Christmas shopping, so the team focused on the quest to buy an item of clothing. To find out what was going through the minds of shoppers during this quest, a list of questions was drawn up, ranging from the religious persuasion of the shopper to the identity of the gift recipient.

Of more than a thousand questionnaires distributed, 366 were returned, and the researchers' first draft of the laws of Christmas shopping received broad support. For example, women did indeed put more effort into rooting around for the ideal gift than men, and more effort was expended by the parents of older children living at home. Not surprisingly, people were found to spend more time and were more willing to consult a clerk when seeking a costly gift such as a mink stole.

There were some interesting exceptions. The survey undermined the suggestion that a companion helps. This is not so, particularly when it comes to a husband who has been dragged along on the expedition. The team concludes, "Women often shop with their husbands, and ask for their opinions, but in truth these wives only want confirmation of their decisions, and do not fully trust their husband's opinions."

Nor do gift lists seem to make much difference when it comes to the amount of time invested in shopping. Even if you are told to buy a sweater for a present, that still leaves plenty of designs and styles to choose from. And the suggestion that there would be a greater investment of time in choosing gifts for "psychologically difficult" recipients, such as the dread mother-in-law, was not backed by the survey.

Other associations emerged, some unexpected. Consumers with more intense religious beliefs tend to rely more on information provided by the store, suggesting that they have a greater desire to come up with the right gift than unbelievers. Those with smaller families tend to seek more help from clerks. Bargain hunters will scour the store for information but avoid salespeople, presumably because they don't trust their advice.

This kind of insight into the Christmas shopping experience can be exploited during the annual spending spree. For example, if stores want to attract men, relatively little money should be spent on packing, signs, and merchandise displays, compared to having salesclerks available, who should be hovering around to help male customers make up their minds. For women, however, the converse is true: they need better labeling and fewer intrusive clerks.

The Concordia researchers are the first to admit that we still have some way to go before we can uncover all the nuances and details of Christmas shopping, but they believe they have taken the first step toward crystallizing those details, in the form of a set of laws.

Lines and Socks

One depressingly familiar Christmas experience is being stuck in a long line at a supermarket, burdened with vast quantities of food and drink. Fortunately for the holiday shopper, we now know more about this misery than ever before, thanks to nine decades of effort by mathematicians studying the behavior of lines.

Although the lines of people in a supermarket are all subject to random delays, on average they tend to move at about the same rate. This feature of the Christmas shopping experience is captured in a mathematical form by the so-called Poisson process, which assumes that people are as likely to arrive at one time as at any other, but that precisely when they arrive is entirely random. This theory has some subtle implications. For example, you might think that to keep supermarket lines down to a minimum, the rate at which checkouts deal with customers should match the rate at which they arrive. "Wrong!" says Robert Matthews of Aston University in England.

Lining theory shows that unless the checkouts process customers faster than they arrive, the lines will just grow and grow — in theory eventually becoming infinitely long. The reason, Matthews explains, is that the checkouts must be able to cope with the randomness in the arrival of customers, which can mean checkouts are empty one minute and backed up with long lines the next. So to keep lines short and customers happy, managers will have to put up with seeing staff doing nothing. Alternatively, managers can slash staffing levels and have customers complain about the length of lines. Alas, managers can't have happy customers *and* constantly occupied staff. One way to deal with this dilemma, often seen at banks, is to set up one line in front of all the registers. The effect of this on waiting times can be dramatic. If, for example, you replace five lines in front of five registers with one line feeding them all, the average waiting time will drop by a factor of five.

Given all this clever theory, why is it that the line next to you so often finishes first? As Matthews points out, even if your line is the same length as both your neighboring lines, all three are equally likely to suffer from random delays, and the chances that yours will suffer the least is just one in three. Therefore, two-thirds of the time, one of your neighbors will finish before you.

Matthews is interested in lines as one of the manifestations of Murphy's Law — "Anything that can go wrong will go wrong" — on which he has published some research in the past few years. He has coined a Murphy's Law of lines. "If your line can be beaten by a neighboring one, it will be."

It is worth taking a short diversion to study another issue that has fascinated Matthews — the seasonal search for socks. An average family of two adults and two children has two pairs of red socks to hang up each year for the nuts, candies, and other gifts from Santa. As Matthews points out, however, this ritual highlights one of the most irritating problems faced in modern life — the odd sock epidemic. Aunts and grandmothers are famous for their propensity to give socks to relatives at Christmas, and here his research can provide useful advice on what sort to buy.

Anyone with a diverse collection of socks has been struck by the proliferation of odd socks in drawers and the hours spent every week in the quest for a complete pair. Fortunately, Matthews has published a solution to this age-old problem in the journal *Mathematics Today.* He resorted to combinatorics, a mathematical discipline used to analyze combinations, arrangements, and patterns, to come up with his solution. After a frenzy of calculation, he uncovered three Murphy's Laws of Odd Socks.

1. If odd socks can be created, they will be. When two socks go missing at random, it's far more likely that they will leave behind two odd socks than conveniently disappear as a pair.
2. Random loss of just half the socks typically cuts the number of complete pairs left by three-quarters — and those that do remain will be lost in a sea of odd socks.
3. Even if you clear out all odd socks, the problem of finding matching pairs remains formidable. You will have to rum-

mage through about one-third of ten pairs of socks to have a reasonable chance of finding a single matching pair.

"At first glance it may seem ludicrous to say that if odd socks can be created, they will be," he says. "However, a moment's reflection reveals its plausibility." Imagine a drawer containing only complete pairs of socks of different designs. The only assumption we need to make for the analysis to proceed is that the loss process is random, with every sock as likely to go missing as any other. If one sock goes missing, it creates an odd sock — the partner of the sock you've just lost. When the next sock goes missing, it could be either that one odd sock just created or a sock from a still-complete pair. As the latter outnumber the former, it is clearly more likely that another complete pair will be broken up, leading to the creation of yet another odd sock. "We can thus see glimmerings of evidence that Murphy's Law really does affect sock drawers," Matthews says.

Attempts to beat Murphy's Laws of Odd Socks usually take the form of practical measures for keeping pairs of socks together, such as putting them into pillowcases before they go into the washing machine. Ideally, of course, we would like to beat the laws without having to go to such lengths.

The simplest solution is to either replace all our distinct pairs of socks with identical ones, go the way of the elderly Albert Einstein and eschew them altogether, or show stoic indifference to odd socks. "Happily, however, combinatoric analysis shows these dreary solutions are unnecessarily draconian. We can allow ourselves a little variety," Matthews says.

He recommends that people in general — or aunts seeking socks as presents in particular — buy two types of socks: the red "Christmas socks" and one other color. Losing half the socks at random typically cuts the number of both types by half, and thus the number of possible pairs by three-quarters, as before. However, the remaining socks are not lost among myriad colors of odd socks, and equal numbers of both types of socks will generally go missing. "And of course you can guarantee that after pulling out just three socks, you can get a matching pair if you are in a hurry in the morning."

Matthews admits, however, that the use of combinatorics has not answered the greatest mystery at all: where do the socks go?

5

FESTIVE FARE

Be nice to yu turkeys dis christmas
Cos turkeys just wanna hav fun
Turkeys are cool, turkeys are wicked
An every turkey has a Mum.

BENJAMIN ZEPHANIAH,
"TALKING TURKEYS"

THE FIRST CHRISTMAS CARD WAS DESIGNED BY the artist John Horsley, an exponent of what the Victorians called the narrative style. The side panels of the card depict acts of charity toward the less fortunate, a familiar theme of the day. However, the subject of the centerpiece, surrounded by leafy trelliswork, is not the traditional Dickensian quest to right social wrongs, but a scene with a great deal of resonance today: a large family enjoying Christmas dinner.

Whether roast turkey, latkes, or stollen, seasonal cooking is nothing more than a branch of applied chemistry, and here science can provide many insights. The color of turkey meat, for example, offers a glimpse of the bird's lifestyle. Scanners traditionally used to investigate the human body can be turned on food to reveal pockets of fat or a sixpence hidden in a plum pudding. Thermodynamics can help us cook that same pudding to perfection. Indeed, scientists have demonstrated that even the brandy sauce owes its consistency to a unique arrangement of molecules, in structures measuring up to fifty-millionths of a meter across.

For all those who worry that these materialist insights somehow jar with the romance of Christmas tradition, don't forget that this tradition has already been subject to very diverse influences. When English-speaking countries settle down for their traditional fare, they eat an Aztec bird by a German tree, followed by a pudding spiced with subtropical preserves. In ancient societies, where hunger was king, the only tradition that held true was that people tended to feast during the long winter days. That continues to be true today. The food, naturally enough, varies from country to country, district to district, year to year, and culture to culture.

Meat of some kind is the most important Christmas dish, often ham (dating back to the sacrifice of a wild boar to a god called Frey during a Scandinavian Yuletide festival) or goose (then turkey), as well as fish such as carp or salmon. Sometimes, the entire lot has been offered. For example, in 1773, in his *Diary of a Country Parson,* James Woodforde wrote of "two fine codds boiled with fryed Souls round them and oyster sauce," served before the beef, pea soup, lamb, wild ducks, "sallad," and mince pies. That evening they ate rabbit.

This spirit of excess is also reflected in Whistlecraft's (John Hookham Frere) rhyme: "They served up salmon, venison, and wild boars / By hundreds, and by dozens, and by scores / Hogsheads of honey, kilderkins of mustard / Muttons, and fatted beeves, and bacon swine / Herons and bitterns, peacocks, swan and bustard / Teal, mallard, pigeons, widgeons, and, in fine / Plum puddings, pancakes, apple-pies, and custard."

Like so many aspects of modern Christmas, the depiction of the festivities as a cozy exhibition of communal gluttony can be found in *A Christmas Carol,* a book written to appeal to the sensual as well as spiritual Victorian sensibilities. "Heaped up on the floor, to form a kind of throne, were turkeys, geese, game, poultry, brawn, great joints of meat, sucking-pigs, long wreaths of sausages, mince-pies, plum-puddings, barrels of oysters, red-hot chestnuts, cherry-cheeked apples, juicy oranges, luscious pears, immense twelfth cakes [frosted cakes eaten on Twelfth Night], and seething bowls of punch."

After reading the book, Thomas Carlyle, not a man given to impetuous gestures, dashed off an order for a large turkey. His wife remarked on how he was seized "with a perfect convulsion of hospitality, and has actually insisted on improvising two dinner parties with only a day between." Indeed, you can scarcely flick through a dozen consecutive pages of anything by Dickens without bumping into a steaming veal pie, succulent oysters, or slabs of beefsteak.

Today a cornucopia of seasonal fare is offered around the world. In Spain a soup of chicken stock, vegetables, and pasta is eaten on Christmas Day. In France *Le reveillon,* a late supper held after Midnight Mass on Christmas Eve, varies according to regional culinary tradition. In Alsace goose is the main course, in Burgundy it is turkey with chestnuts, and in Paris people often feast on oysters and pâté de foie gras.

In Roman Catholic countries fish plays a part, notably on the holy fasting day of Christmas Eve. On Christmas Day stuffed baked carp in a rich sauce is often eaten in Czechoslovakia and other places where the fish can be caught. Pickled fish is served in Poland, eels and squid appear in some Italian towns, and the Swedes prepare a remarkable dish called lutefisk — fish that has been soaked in lye. The alkaline water softens the tissue by dissolving its proteins.

The diversity of Christmas cooking also reflects the availability of food at that time of year. In Paris, during the Prussian siege of 1870, the zoo and the sewers provided the Christmas dinner at Voisin's, a fashionable restaurant: consommé of elephant, braised kangaroo, antelope pâté, and a whole cat garnished with rats.

Today many Christmas revelers in the dry interior of southern Africa will tuck into the local harvest of mopane worms — fat, spiny, mottled caterpillars of the emperor moth (*Gonimbrasia belina*). Recent research has found that in terms of protein, fat, vitamins, and calories, the caterpillars compare favorably with meat and fish.

There are, of course, many dishes associated with the other celebrations that take place in December. For example, Kwanzaa features a variety of fruit and corn, while Hanukkah means latkes (potato pancakes) and sufganiyot (special doughnuts fried in oil).

The Turkeys Invade

For a long time in Germany, Britain, and elsewhere, the goose was the most popular Christmas dish. Then came the turkey. The Milanese Girolamo Benzoni, who traveled in Central America in the sixteenth century, remarked in his *History of the New World* (1565) that these "Indian fowls" were one of the unique contributions made by the area.

This outlandish bird started its invasion of Europe around 1519, when ships first brought turkeys back to Spain. It marched on to the Spanish Netherlands and then to East Anglia, in England, where turkey farms were established. The eighteenth century saw great turkey drives from Norfolk to London. On Christmas Day 1815, the English essayist Charles Lamb wrote of "the savoury grand Norfolcian holocaust."

In 1851 the turkey replaced the swan as the Christmas bird at Queen Victoria's table. It was not until late Victorian times, however, that the turkey superseded the goose, or in the north roast beef, as the leading Christmas fare in England. (In America, of course, wild turkey, being an indigenous bird, was a popular festive dish earlier.)

By 1900 *The Royal Magazine* estimated that in London and its surroundings alone, "the turkeys and geese cooked for Christmas would form an army, marching ten abreast, which would reach from London to Brighton," and the champagne used to wash them down would "keep the Trafalgar Square fountains working incessantly for five days." Once families had consumed what they could of the roast turkey, there followed cold turkey — lots of it — minced and curried and deviled for days and days to come.

Today the traditional turkey is likely to be altered by genetic engineering. Bernie Wentworth and his colleagues at the University of Wisconsin at Madison say that turkeys could lay 20 percent more eggs if they did not lapse into periods of broodiness. The aim of current research is to redesign the birds so that they produce very little prolactin, the hormone that triggers broody behavior. Wentworth has made an "antisense" fragment of DNA that sticks to the gene responsible for prolactin, preventing production of the hormone.

Biofundamentalists claim that such work will, in effect, reduce the creatures to machines that perform some market function. Whatever the ethics of designing creatures for human use, this is but one application of genetic engineering, which is certain to make a considerable contribution to the taste and texture of future Christmas meals by mixing and matching genes from a wide range of species, from bacteria and fungi to pigs and fish.

The Turkey Revealed

Whatever Mrs. Beeton in Britain or Martha Stewart in America may see in a turkey, a chemist regards the bird as merely a combination of water, fat, and protein, in proportions of around 60:20:20. Most of its meat is muscle tissue and consists largely of proteins. Two workhorse proteins — myosin and actin — make up the fibers that give muscle its texture or grain. They lie in layers and slide past each other when the muscle is stimulated to contract.

The color of turkey meat holds a record of the bird's lifestyle. The difference between the white and dark meat is not due to its blood, or the red oxygen-carrying hemoglobin, but to the closely related oxygen-*storing* myoglobin, a molecule that also contains an iron atom at its heart. Myoglobin retains and stores the oxygen brought by blood until muscle cells need it. To some extent, the oxygen demand can be related to the general level of activity: muscles that are exercised often and strenuously need more oxygen.

Two general classes of muscle fibers are at work during exercise: fast and slow. Fast, or white, muscle fibers are used for rapid motion and rely on blood for their energy. Slow, or red, muscle fibers are used for more sustained motion, fueled by stores of fat. And that is where the red, oxygenated myoglobin comes in. The fast fibers are white because they do not need this red oxygen-storage molecule.

Turkeys do a great deal of standing around but little if any flying, so their breast muscle is white and their legs are dark. Leg meat also is greasier than breast meat because fat is required to fuel the red muscle fibers. By contrast, game birds spend more time on the wing, and their breast meat may be as dark as their drumsticks, seasoned with myoglobin throughout.

Molecular makeup also affects the texture of the meat. Turkey legs are tougher than breasts because they are used more. As the bird grows and exercises, the muscles enlarge by increasing the number of actomyosin filaments within the muscle fibers. The more filaments there are to cut through, the tougher the meat. In general, however, well-exercised tough meat is tastier than tenderer, less-exercised meat.

The tenderness and appearance of cooked meat depend on the reaction of its component chemicals. At the molecular level, that means they depend on how the various molecular components respond to the shake, rattle, and roll of heat energy.

When myoglobin is heated during cooking, it turns brown because the iron at its center is oxidized, changing the way the molecule absorbs light. In the presence of heat, proteins and other components break down into fragments. When these are small and volatile they add to our perception of flavor. Some originate from a molecule called adenosine triphosphate, or ATP, the chemical energy currency of cells. Of the vast cocktail of chemicals produced by cooking, the two most responsible for a meaty flavor are inosine monophosphate, which arises from the decomposition of

ATP, and monosodium glutamate, the sodium salt of a naturally occurring amino acid.

When you roast a turkey, the muscle fibers contract, until at temperatures above about 80°C (about 180°F) individual cells within the fibers begin to break up. At the same time, roasting snaps the bonds within molecules (hydrogen bonds and so-called disulfide bridges) that maintain the shape of the proteins, which are coiled up chains of amino acids.

As the proteins begin to unravel, they adopt more open conformations, so that the meat becomes tender. But if you continue roasting for too long, the proteins will build a network of new chemical bonds. This cross-linking of proteins replaces the delicate association that existed before, making the meat tough in a process called coagulation.

Another cause of toughness is the connective tissue that attaches the muscles to the bones. This contains three more proteins — collagen, reticulin, and elastin. Neither reticulin nor elastin are much weakened by the heat of cooking, but the three strands that make up collagen unwind into separate ones, better known as gelatin, which is soft. The secret of an excellent roast, then, is to denature the collagenous tissues, while at the same time avoiding too much coagulation of the muscle proteins.

Calculating the time required to cook a turkey demands skill. Denaturing and coagulation of proteins take place at different temperatures and rates, depending on the proteins' location in the turkey. Fortunately, you can rely on a handful of general principles, according to Peter Barham, a physicist at the University of Bristol, England. The longer the bird remains at a high temperature, the more moisture it will lose, and the greater chance there will be that muscle proteins will coagulate. A minimum temperature of 70°C (about 160°F) will convert collagen to gelatin and ensure that the muscle fibers break down. If you want your bird to be both tender and cooked, it would be wise not to exceed this temperature. Thus, at its simplest, the optimum cooking time can be defined as the minimum time required to heat the center of the bird to 70°C.

Turkey Thermodynamics

The next problem — calculating how long the bird should stay in the oven — is one for the science of thermodynamics. This discipline took shape with the advent of steam power in the Industrial Revolution in Britain during the early nineteenth century. The word is formed from the Greek words meaning movement and heat, and its principles can be just as easily applied to turkeys as to steam engines.

In 1947 H. S. Carslaw and J. C. Jaeger went some way toward solving this seasonal problem in a reference work entitled *Conduction of Heat in Solids*. They came up with equations that govern heat transfer in a uniform sphere. Little did they know that they had provided an excellent starting point for a theory to find the correlation between the radius of a turkey and its cooking time.

To ensure that the mathematics did not get out of hand, they made a few simplifying assumptions. In the Christmas context these can be described as follows. First, the oven must be maintained at a constant temperature throughout. Second, the thermal diffusivity — a measure of how fast heat passes through the turkey — is independent of temperature and time. Finally, and most important, the turkey must be extremely plump, so much so as to be spherical.

The time for heat to diffuse through the tissue so that the

center of this spherical turkey reaches a certain temperature is proportional to the square of the radius of the turkey. Then we have to figure out the size of the idealized bird. By assuming that the spherical turkey has the same mass, *M*, as the real one, we can calculate the radius of the required spherical turkey. To do this we can resort to a well-known formula that says the mass of a sphere is proportional to the cube of its radius.

By combining these two formulas, the radius disappears from the calculation, and we are left with a cooking time, *t*, proportional to *M* to the power of two-thirds. This would, of course, hold true only if the turkey was a perfect sphere. Barham points out that a cookbook's simple algorithm of twenty minutes per pound plus twenty more minutes for a small turkey, or fifteen minutes per pound plus fifteen minutes for a large bird, is an equally good approximation for a real turkey.

And then there are all the trimmings to prepare. You do not need any fancy mathematics to determine that when you stuff a turkey, it is better to put sausage meat or whatever into the neck, where it will reach a higher temperature than in the middle of the bird, where it might be undercooked.

The vegetables that accompany the turkey require us to think again about how molecular structure reacts to cooking. Potatoes contain granules of starch — a carbohydrate consisting of long chains of sugar molecules — which become soft, swollen, and gel-like when cooked in water heated to between 58° and 66°C (approximately 135° to 150°F). At this point the granules begin to soak up water, and they swell to many times their normal size. The perfect potato is full of such swollen, tender granules.

Hanukkah latkes are dollops of grated potatoes mixed

with eggs, matzo meal, and salt that are fried in a well-oiled pan. As well as having symbolic significance (recalling the jar of oil that miraculously burned for eight days and eight nights when Judas Maccabees rededicated a temple that had been desecrated by the Syrian tyrant Antiochus more than two thousand years ago), the oil has several important roles to play: it ensures that the mixture is in uniform contact with the heat source; it lubricates and prevents any sticking to the pan; and it supplies some flavor by enabling browning reactions to occur.

During these chemical reactions, a carbohydrate unit from the potato starch reacts with the nitrogen-containing amine group on an amino acid, which may be free or part of a nearby protein molecule. An unstable intermediate structure is formed and then undergoes further changes: this is called the Maillard reaction and brown coloration and intense flavor result from its by-products.

The trick is to brown the outside of the latke and ensure that the inside is done, since the moist interior remains largely water and never exceeds the boiling point. Thus the potato has to be grated to ensure that there is no serious disparity between superficial and deep latke cooking times.

Roast potatoes also present a challenge. When the British traditional roast dinner emerges, the potatoes are often cause for disappointment, whether because of their anemic appearance or because they look cooked on the outside but are hard inside. Too high a temperature will brown the surface before the interior is cooked, and too low a temperature will cook the interior without browning the exterior.

The roast potato problem can be tackled by a two-stage cooking process. First the potatoes should be boiled for a few minutes so that they acquire a gel-like surface layer. Then they should be basted with fat and put in the oven. During roasting this surface layer prevents the starch granules underneath from absorbing too much fat, while the surface layer reaches about 160°C (320°K). The starch on the surface degrades and oxidizes to give the characteristic crispy brown coating.

Green vegetables require a quite different cooking philosophy from starchy ones such as potatoes and winter squashes. They require the briefest weakening of their cell walls by heat and the extraction of a little water to make them tender. The bright green that some vegetables develop a few seconds after being thrown into boiling water is a result of the escape of gases trapped in the spaces between their cells. The departure of these gases reveals the pigment (chlorophyll) that plants use to harvest sunlight. With a magnesium atom at its heart, the chlorophyll molecule absorbs both violet and red light, so that reflected light appears green. If you continue cooking the vegetables, however, this magnesium atom will be replaced by charged hydrogen atoms (protons), dulling the color to the insipid green we all associate with overcooked brussels sprouts.

The Dreaded Brussels Sprout

The brussels sprout, a member of the cabbage group, was thought to have been developed in northern Europe in about the fifth century, although the first clear record of the vegetable dates from much later, in 1587. This vegetable is now part of many traditional Christmas meals, yet, paradoxically, it is greatly disliked. Few people think to ask why it is so hated by so many, particularly children, and herein lies a fascinating scientific tale. The sprout's slightly bitter, sulfurous taste is meant to discourage would-be insect diners. This is the vegetable equivalent of a chemical weapon and evolved to deal with pests. Yet we are not hurt in the crossfire. (Although there is some unpleasant fallout as stomach bacteria do convert the sulfur into foul-smelling hydrogen sulfide.) Indeed, sprouts and their close relatives are among the most nutritious leafy vegetables, rich in minerals, fiber, protein, carotene, and vitamin C. These plants, as well as many other fruits and vegetables, also contain a range of non-nutrient compounds that seem to protect against a wide range of cancers, including breast, lung, and colon cancers, by reducing damage to cell DNA.

Brussels sprouts are rich in one family of secondary compounds (called glucosinolates, notably sinigrin). Ian Johnson of the Institute of Food Research in Norwich, England, has found that sinigrin suppresses, at least under laboratory conditions, the development of damaged cells that may eventually develop into full-blown tumors. The protective chemical seems to be a breakdown product of sinigrin, called allyl isothiocyanate, a volatile molecule that is largely responsible for the smell and taste of sprouts.

How is it that we not only eat these natural pesticides with impunity but also — hydrogen sulfide apart — benefit from the experience? The answer may lie in our evolutionary past. Thanks to our long history as omnivores, consuming a panoply of fruits and vegetables, we find ourselves today biologically adapted to an intake of plant toxins. We seem to be so well adapted that our health suffers if we are deprived of a steady supply of the plants' chemical weapons.

The aversion of children to such vegetables also may have to do with our Stone Age ancestry. Our bodies — and our appetites — have evolved as a direct result of survival of the fittest, an endless arms race between our ancestors over millions of years as they fought off parasites, predators, and disease. Physiologically, human beings still belong in the Stone Age, when the diets of our hunter-gatherer ancestors were very different from that of today, notably in terms of fat and salt content. From the Darwinian perspective, an aversion to bitterness may have evolved to help animals avoid poisonous plants, since many natural products that taste this way are harmful. For example, alkaloids, such as strychnine, are poisonous, and the ability to detect them by taste was a necessary adaptation for survival. We can, of course, train ourselves to like bitter tastes, and sometimes for good reason. The bitter ingredients in aperitifs, such as the quinine in a gin and tonic, stimulate the production of saliva, making them the ideal prelude to a meal.

In early childhood, then, we all tend to dislike bitter tastes, a preference that helped protect our ancestors. George Williams, professor emeritus of the State University of New York and one of the pioneers of Darwinian medicine, suggests that children's feeding instincts are to look for the blandest foods and to avoid strong flavors because these flavors are linked to high levels of toxicity. At this critical stage in their development, children are adapted to avoid toxins to a greater extent than adults.

Children also have a preference for sweetness, again reflecting their innate programming to avoid poisons, but also ensuring that they will adore breast milk, according to Dave Mela of the Institute of Food Research in Reading, England. Moreover, a taste for sweetness enables a person to tell when fruit is ripe.

The nutritional value of vegetables is another reason why

children avoid them instinctively. Vegetables are not particularly energy dense, Mela says, and calories were a crucial consideration of our ancient ancestors, given the lack of four-wheeled transportation and the constant need to hunt food, fend off predators, and so on. "That is an evolutionary adaptive response that argues against a liking for vegetables, which are not sweet or energy dense."

This childhood aversion has serious consequences. A recent report from the Cancer Research Campaign in Britain, presented by its director general, Gordon McVie, claims that Christmas dinner is now the only meal each year where most British children get an adequate intake of vegetables. According to the report, "Of the 300,000 people who get cancer each year in the UK, about a third of cases are diet-related and potentially preventable. Getting children to eat vegetables is not a recent problem facing parents. Repeated generations have tried to cope — often with limited success. Much needs to be done to encourage children to eat more vegetables."

There is an intriguing complication to the brussels sprout's story. Some people find the taste of sprouts much more bitter than others. This might not influence their food preferences as adults, since most of us learn to love bitter flavors such as those found in coffee, tea, and beer. As children, however, these so-called supertasters may have a particular aversion to bitter vegetables — not just sprouts but broccoli, too.

A pioneer of supertaster studies, Linda Bartoshuk of the Yale University School of Medicine, believes that one-quarter of all Caucasians are supertasters, one-quarter are nontasters, and the rest are mediumtasters. Each group lives in a different taste world, so that to supertasters sugary Christmas confections taste sweeter, caffeine seems more bitter, and the burn of a chili pepper may be twice as intense as that experienced by nontasters (who have little or no taste for certain substances).

Supertasters, mostly women, are usually distinguished because they find the chemical 6-n-propylthiouracil so bitter that they cannot stand it. They can also be identified by examining the little projections on the tongue that contain the taste buds. "When someone sticks out his or her tongue, we can tell just by eyeballing it," says Laurie Lucchina, a member of the team at Yale. Supertasters can have up to thirty projections in a six-millimeter-diameter circle, while nontasters can have as few as ten projections in the same area, and they tend to be much larger in supertasters.

The phenomenon is genetic. Because of the way the supertaster trait is inherited, if mediumtasters marry, one-quarter of their children will be nontasters, one-quarter supertasters, and the rest mediumtasters. Mediumtaster parents of a large family may not realize that this is why only some of their offspring have a particular aversion to sprouts. The same trait is also a cause for concern in the fight against cancer. "Many antioxidant flavonoids that are so important for cancer prevention are either bitter or occur in bitter-tasting vegetables and fruits," says Adam Drewnowski of the University of Michigan. For example, supertasters dislike naringin, an antioxidant that is the principal bitter ingredient in grapefruit juice, which is being examined for its value in inhibiting cancer. The health benefits of bitter chemicals creates a dilemma for producers and genetic engineers. If they breed low levels of sinigrin into the plants to produce milder-tasting brussels sprouts, for example, they risk reducing the health benefits and perhaps making the plants more vulnerable to certain pests as well.

Scientists and Puddings

Plum pudding is a food fit for a monarch. King George I, sometimes called the "Pudding King," tucked into one at six o'clock on December 25, 1714, his first Christmas in England. We still have the recipe: 5 pounds finely shredded suet; 1 pound eggs; 1 pound dried plums, stoned and halved; mixed peel, cut in long strips; small raisins; sultanas; currants; sifted flour; sugar; brown bread crumbs; 1 teaspoon mixed spice; ½ grated nutmeg; 2 teaspoons salt; ½ pint "new milk"; juice of ½ lemon; "a very large wineglassful brandy."

This boiled confection also makes a star appearance in *A Christmas Carol:*

> The pudding was out of the copper. A smell like a washing-day! That was the cloth. A smell like an eating-house and a pastrycook's next door to each other, with a laundress's next door to that! That was the pudding! In half a minute, Mrs. Cratchit entered — flushed, but smiling proudly — with the pudding, like a speckled cannon-ball, so hard and firm, blazing in half of half-a-quartern of ignited brandy, and bedight with Christmas holly stuck into the top. Oh, a wonderful pudding! Bob Cratchit said.

Christmas pudding had its origins in frumenty, a type of porridge made from hulled wheat spiced and boiled in milk, which was sometimes used as a fasting dish on Christmas Eve or as an accompaniment to meat. Over the years more ingredients were added: eggs, mace, and prunes. Indeed, meat was often thrown in to make a plumb porridge.

Some claim that this dish merged with a great boiled sausage called a hackin, an ancestor of that culinary target of many jokes about Scotland, the haggis. Somewhere along the way, the hackin disappeared or became one with the pudding, which became more solid and was boiled in a cloth. The result was the pudding of the kind that Mrs. Cratchit unveiled: a skinless relative of a haggis that is set alight, is garnished with evergreens, and sometimes contains coins, a practice that some argue is a lingering reminder of a Saturnalian tradition involving the drawing of lots. Today, scientists are perhaps the most determined consumers of plum puddings. Here, I am thinking of the puddings that are sent by the British Antarctic Survey to its teams in Antarctica. The team members can sometimes find themselves up to 1,200 miles from the nearest base on Christmas Day.

The best puddings can take an age to prepare. If they are left a year or two to mature, slow chemical processes similar to those in a bottle of wine in a cellar take place. Then the pudding must be cooked. Like the roasting of the turkey earlier in this chapter, the thermodynamics of cooking Christmas pudding has been studied by physicists. Glenn Cox, a physics lecturer at the University of Birmingham, England, is surprised to find that cookbooks rely on rules of thumb rather than deferring to the scientific principles of thermodynamics to work out how long a dish should stay in the oven and at what temperature.

Cox derived an equation to describe the time required to cook a Christmas pudding — that is, for the center of the pudding to reach a given temperature. Cooking up the equation was relatively simple. As for turkey, it was necessary to make a number of simplifications so that his mathematical recipe was manageable: the pudding had to be spherical and homogeneous (a fancy way of saying that currants, threepenny bits, and so on were not taken into account). Cox also ignored the way heat-driven chemical reactions in the pudding affect the way it is cooked. Finally came his trickiest task — finding the thermal diffusivity of the pudding, the factor that describes the pudding's ability to store and transmit heat. Cox used a guesstimate based on a comparison with substances such as salt, sand, beeswax, charcoal, water, and alcohol.

To derive his pudding equation, he resorted to complicated thermodynamic and mathematical equations, such as spherical Bessel functions and Fourier series. But what his equation does is simple: essentially it describes the temperature distribution in a sphere when its surface is held at a constant temperature. His work, which is similar to that on the turkey, revealed that the cooking time should vary in proportion to the square of the radius of the pudding, or as the volume to the power of two-thirds.

However, unlike with the turkey formula, many cookbooks do not agree with his predictions. Cox was surprised to find that many make no distinction between the cooking times for one-liter and two-liter puddings. Yet his simple analysis of pudding thermodynamics showed that one should expect to cook a two-liter pudding approximately 1.6 times longer than one half the size. Of course, Cox is the first to admit that even a properly cooked pudding can taste awful. "The proof of the pudding is in the heating," he observed — with a surprisingly straight face.

Christmas Sauces

One traditional way to cope with an unappealing pudding is to smother it with a sauce. Here we find another rich seam of Christmas science. These liquids are of huge interest to surface scientists because of their complicated physical structures. Cream, brandy butter, and brandy sauce owe their properties to a unique arrangement of molecules, in structures measuring up to fifty-millionths of a meter.

All these toppings are colloids — suspensions of various substances dispersed within each other — and are important to the food industry because they affect the taste and texture of foods and how long they keep. An emulsion is a colloid composed of two immiscible liquids, where one liquid, usually oil or fat, is dispersed as droplets within another, usually water. Ice cream contains these two components, as well as solid fat particles, ice crystals, and tiny air bubbles. It is the ultimate colloid, because it contains solid, liquid, and gas.

Cream consists of droplets of dairy fat coated with a protein (casein) that are dispersed in a watery solution of whey proteins. The concentration of the oil droplets dictates the creaminess, with cream having a higher concentration than milk.

Brandy butter has a structure like that of cream that has been turned inside out: droplets of water are suspended in oil. Experiments conducted by Margaret Robins and Mary Parker at the Institute of Food Research in Norwich, England, suggest that it is not possible to get the brandy into the suspended water droplets in butter without adding sugar. Through a process that is not yet understood, the sugar helps the brandy pass through the surrounding fat layer.

Brandy sauce has a more complex structure. Like cream, it can be thought of as a suspension of materials in a watery solution. In this case, however, it owes its thickness to starch polymers — long chains of sugar molecules — that are packed together in flour granules.

The starch polymers come in two forms. One type, called amylose, consists of chains. The other type, called amylopectin, consists of branched molecules. Both types cling together within flour granules by means of hydrogen bonds (see Chapter 8). When water gets hot enough, its component molecules have enough energy to penetrate the granules, so the molecular structure begins to fall apart. As a result, these polymers are released into the watery solution of milk proteins, sugar, brandy, and other components of brandy sauce.

The starch molecules cling to the water molecules and form an entangled mesh that captures the water-swollen granules. This dispersion of a solid material (starch) in a liquid (water) is known as a sol and is what makes the brandy sauce thicken. However, boiling, vigorous stirring, or heating for a long time will thin the sauce, because the granules will eventually disintegrate to leave a soup of finely dispersed starch fragments.

Chocolate

Christmas is a time for eating chocolate — lots of it. Consumption has come a long way since the first "eating" chocolate was introduced to England by the Bristol firm of Fry and Sons in 1847. Much debate and mythology surround people's craving for this confection, which has been blamed on depression, the menstrual cycle, sensory gratification, or some of the three-hundred-plus chemicals that it contains.

The sensuous properties of chocolate depend on the fat it contains. Cocoa butter can solidify in half a dozen different forms, each of which has a different effect on "mouthfeel" and palatability. Form V predominates in the best chocolate, making it glossy and melt in the mouth.

Unlike other plant edible fats, which are usually oils, cocoa butter is enriched in saturated fatty acids so that it is solid under normal conditions and has a sharp melting point of around 34°C, just below body temperature. Heat is absorbed when this occurs, giving a sensation of coolness on the tongue.

Another reason we like chocolate is the stimulatory effects of caffeine and related chemicals. Every 100 grams of chocolate contains 5 milligrams of methylxanthine and 160 milligrams of theobromine (named after the cocoa tree, whose botanical name, *Theobroma cocoa,* means "food of the gods"). Both are caffeinelike substances.

Originally, chocolate was a stimulating drink. The name is derived from the Aztec word *xocalatl,* meaning "bitter water." In the seventeenth century a physician from Peru wrote how it is "good for soldiers who are on guard." Indeed, some people have suggested that it was Casanova's favorite bedtime drink — to give him a boost when he needed it. Medical textbooks do note, however, that when taken in large quantities, these stimulants can induce nausea and vomiting. This effect can also be observed in children (and others) who overindulge on Christmas Day.

Every 100 grams of chocolate also contains 660 milligrams of phenylethylamine, a chemical relative of amphetamines, which has been shown to produce a feeling of well-being and alertness. This may be why some people binge on the stuff after an upsetting experience — or perhaps to cope with the stress of Christmas shopping.

Phenylethylamine may trigger the release of dopamine, a messenger chemical in the brain that plays a role in the "reward pathway" that governs our urge to eat or have sex. Phenylethylamine raises blood pressure and heart rate, and heightens sensation and blood glucose levels, leading to the suggestion that chocoholics "self-medicate" because they have a faulty mechanism for controlling the body's level of the substance. However, if a person consumes too much phenylethylamine or has an inability to remove it due to the lack of a key enzyme (monoamine oxidase), blood vessels in the brain constrict, causing a migraine.

More recently, it has been found that chocolate also contains substances that can act like cannabis on the brain, intensifying its other pleasurable effects. Three substances from the N-acylethanolamine group of chemicals can mimic the euphoric effects of cannabis, according to a study by Daniele Piomelli, Emmanuelle di Tomaso, and Massimiliano Beltramo of the Neurosciences Institute in San Diego. Their work dates back to 1990, when scientists found a site in the brain that responds to cannabinoids, the class of compounds that includes the active ingredient in cannabis. Recently they have discovered the specific substances in the brain that bind to this site. One is a fatty molecule dubbed anandamide after the Sanskrit word for "bliss." Piomelli investigated chocolate, which is rich in fat, because he correctly suspected that it might contain lipids related to anandamide. Piomelli was first inspired to look into the mood-altering effects of chocolate when he became addicted to the stuff one gray winter in Paris. Now that he has moved to California, which is as sunny as his homeland of Italy, he is no longer a chocoholic.

Psychological theories have also been advanced to explain the attractions of chocolate. David Booth, a psychologist and nutritionist at the University of Birmingham, England, concedes that some of chocolate's chemical constituents may produce slight side effects but attributes our obsession with it to emotional and social conditioning early in life, when we turn to it for comfort. Another psychological spin on Christmas chocolate consumption comes from Peter Rogers of the Institute of Food Research in Reading, England, who believes that we crave chocolate largely because social norms say that we should eat it with restraint.

Fighting the temptation to eat lots of chocolate at Christmas may be good for your soul but may undercut your resolve. Indeed, it may even be dangerous, according to a fascinating experiment by Ellen Bratslavsky and Roy Baumeister at Case Western Reserve University. They tested students for self-control, using one plate of chocolate cookies and another of radishes. In the test, some students were told to eat radishes and not cookies, while others were told to eat cookies and not radishes, before tackling an insoluble puzzle. Students who resisted the temptation to eat the forbidden cookies worked on the puzzle for an average of eight minutes before giving up. Those whose self-control had not been tested persisted for twenty minutes. "When people exert self-control — such as when under stress — self-control may fail in other spheres. Resisting temptation can be hazardous to your mental health," Bratslavsky says.

The Smell of Christmas

We associate many smells with Christmas. Intriguingly, the effect of these aromas is more subtle than you may think. Among several food smells tested by Neil Martin, a psychologist at Middlesex University in Enfield, England, chocolate was found to exert a significant calming effect on the brain.

Martin asked forty volunteers to sit in a "low-odor room," wearing goggles and headphones to block out other stimuli, while he wafted smells their way in two experiments. All the while he used electroencephalography (EEG) to record their brain waves as they sniffed. Subjects were first exposed to the odors of real food: rotting pork, chocolate, coffee, and baked beans. They then sniffed synthetic food aromas: almond, garlic and onion, strawberry, spearmint, vegetables, cumin, and chocolate. Apart from chocolate, the smells had little effect on the subjects' theta brain waves, which are associated with attentiveness. Only chocolate reduced brain activity, a finding that may be related to the odor's ability to relax sniffers and make them inattentive. Chocolate connoisseurs might not be too surprised by this discovery.

Smells can even help us carry out tasks, according to an experiment by Martin's colleague Alison Gould. In one experiment, subjects completed a tedious "visual vigilance task" in the presence of either no odor, an alerting (peppermint) smell, or a relaxing (bergamot) one. They did better with the peppermint. Another study showed that subjects performed an unchallenging task better when stimulated by an unpleasant smell — in this case, sour milk. A pleasant smell, that of an air freshener, helped them perform complex tasks better.

This kind of work is part of a growing body of evidence

suggesting that the rich smells of Christmas may affect us more than we realize. We become aware of different smells because inhaled air carries odor molecules, from the pyrazines in chocolate to cinnamon's cinnamaldehyde, to the roof of the nose. This area is covered with membrane containing sensitive, hairlike nerve fibers — the sensory receptors of the olfactory nerves, which pass directly into the olfactory "bulb" in the brain. As this bulb lies close to the brain center responsible for emotions and cognitive behavior, it is not surprising that smell is closely linked with emotions.

A smell can transport people back in time, evoking a memory such as a childhood Christmas. This kind of effect was observed by Howard Ehrlichman and Jack Halpern of City University of New York, who asked individuals what memories were evoked by neutral words. Happier memories emerged when the individuals were exposed to a pleasant odor (almond) than to an unpleasant one (pyridine).

The intimate link between the smells of Christmas — whether of roast turkey, latkes, or a spiced cake from Nuremberg's famous winter fair — and our emotions was recently revealed in experiments conducted by David Zald, a research fellow at the Veterans Affairs Medical Center in Minneapolis and at the University of Minnesota.

Working with José Pardo, Zald studied women exposed to a variety of smells while inside a brain scanner. He chose women because they tend to find smells more intense than men do. For the study, a dozen women in the scanner were exposed to the smells from plastic bags. Moderately bad smells included garlic breath, natural gas, and motor oil. The worst stench was the sulfur-bearing odor similar to rotten eggs or a sewer.

When volunteers sniffed this, a pair of almond-shaped

complexes deep in their brains kicked into overdrive. Each half of the brain has one of these cell clumps called amygdalae, and together they're a key part of the brain's machinery for creating emotional reactions. Scientists have long known that the link between smells and emotions shows up anatomically: there's a direct connection between the amygdalae and the brain machinery that processes information from the nose. In the case of a bad smell, they can see this connection being used, as if the amygdalae are telling the rest of the brain, "Hey, you really hate this stuff." Pleasant study smells, which included fruits, flowers, and spices (and presumably would include Christmassy smells), evoked a weak response, and only in the right amygdala.

The context of the smell is very important, Pardo notes. If, for example, the brain is supposed to evaluate the smell of wood burning and the person is enjoying a cozy fire in a fireplace, "probably the amygdala codes it as, 'This is good, you can enjoy this,' whereas if you're in the middle of a dark theater and you smell smoke, it's 'fear, terrible, get out.'"

One of the women studied in the experiment provided an example of the influence of context. She found that a really bad odor wasn't too terrible, and the brain scan showed that her amygdalae agreed. The reason: she had spent a summer vacation in Alaska near an oil refinery. "It reminded her of that wonderful summer she had," Pardo says.

The link between smell and emotion has been underlined by a study by Alan Hirsch, an odor expert at the Smell and Taste Research Foundation in Chicago. He found that male sexual arousal is strongly linked to food aromas. This find becomes somewhat perplexing when you consider that pizza and popcorn are more of a turn-on than the finest perfumes. "The floral perfumes that we tested showed a median 3 percent increase in penile blood flow; cheese pizza caused a 5 percent increase; buttered popcorn caused a 9 percent increase; and a combination of lavender and pumpkin pie caused a 40 percent increase," Hirsch notes.

He even believes that smells can help you lose weight. "The smell and taste senses work hand in glove. By saturating the sense of smell, you can fool the brain into thinking you've just eaten. You might think that giving people tempting smells would make them eat more. In fact, the reverse is true."

This could come in handy after the Christmas bingeing, if one study is any indication. More than three thousand overweight volunteers were given three vials, each containing a different food fragrance, to sniff each time they became hungry. They were advised to make no conscious changes to their diet or exercise. On average, the volunteers lost five pounds a month for six months. The food odors tested for weight loss included banana, peppermint, green apple, cranberry, baked bread, barbecued meat, vanilla, and cola.

After the Feast

If you happen to celebrate with a midday meal for Christmas, Hanukkah, or Kwanzaa, you might well find yourself snoozing in the afternoon. (In Britain, this is when the monarch's speech is broadcast, reinforcing the soporific effect of the meal.) The reason we tend to feel very sleepy after a heavy meal has been studied by many scientists, including Jim Horne, director of the Sleep Research Laboratory at Loughborough University, England. "We humans are designed to sleep twice a day — once at night and a short nap in the early afternoon — but in this part of the world we tend to repress that," Horne says. The nap is a remnant of the same primeval programming that makes all animals in the bush rest in the hot afternoon sun to avoid the heat. Hot environments reinforce that need, and many cultures living near the equator "have conceded to the inevitable, where the afternoon siesta [a corrupted form of the Latin *sexta hora,* meaning the sixth hour after wakefulness] is a way of life," he says.

Horne studied the contribution of alcohol to the sleepiness. "The theory is that if you are more sleepy in the early afternoon, then it figures that alcohol will be more potent then. One would figure, then, that a pint of beer at lunchtime has more effect than in the evening, when people are more alert. Indeed, we find that it has about twice the effect."

Alcohol interacts with the circadian rhythm of sleep to enhance afternoon sleepiness, so that one pint of beer at lunchtime is equivalent to a quart in the evening. All of this carries a serious implication. "Drivers ought not to drink at all at lunchtime, and the legal blood alcohol limit is no guide to 'safe' driving here," Horne says. For those who wish to enjoy the queen's Christmas speech or feel invigorated after a heavy lunch of any kind, Horne recommends a catnap. But this nap should be less than fifteen minutes, or, as Horne warns, "sleep really sets in, and one can wake up feeling very groggy and far sleepier than to begin with." Alternatively, mild exercise, a blast of cold air, a splash of cold water on the face or, better still, a strong cup of coffee can all help you to keep alert for the remaining Christmas festivities.

6

GLUTTONY: SANTA'S GENETICS

And so I awaited Christmas Eve, and the always exciting advent of fat Santa. Of course, I had never seen a weighted, jangling, belly-swollen giant flop down a chimney and gaily dispense his largesse under a Christmas tree.

TRUMAN CAPOTE, "ONE CHRISTMAS"

THINK OF THE WAY SANTA IS USUALLY DEPICTED on Christmas cards. One aspect of his appearance is obvious, yet attracts little comment. Not his white beard, ruddy complexion, or tendency to chuckle at any opportunity. Nor do I mean the company that he keeps: reindeer, snowmen, and his diminutive helpers. I am thinking of another common Santa trait, one that — amazingly — gets little attention, despite our image-conscious society. It is that huge stomach. Generations of children have asked how he manages to squeeze down chimneys. But few if any have asked the more obvious question: why is Santa so fat? After all, if he lost a few pounds, surely his job would be that much easier. A thinned-down Santa also would provide a role model for moderation and self-control during the holidays.

Perhaps Christmas card artists pay homage to seasonal excesses and Santa's overwhelming cheerfulness by equating rolls of flesh with peals of laughter. Perhaps his girth is the result of eating the millions of cookies and mince pies left out for him on Christmas Eve.

But what if the familiar rotund Santa was more than just the cliché of the jolly fat fellow and generations of artists knew something we did not? What follows is, I admit, speculative. But recent advances in genetic science have led me to suspect that our favorite Christmas icon has a defect in a particular gene. Because a gene is disrupted by a "spelling error" in his DNA, Santa has a propensity to pile on the pounds. Indeed, there is strong evidence that he also may suffer from diabetes.

Santa is not alone. Obesity is now the most common nutritional disorder in the Western world. In America, for example, the epidemic is well under way. An estimated one-third of American adults are overweight, and the incidence of obesity is rising particularly fast in children. Since 1976 the prevalence of pediatric obesity has increased more than 50 percent. Eight out of ten obese adolescents grow up to be obese adults.

In Britain one-third of all people carry too much fat, and about 5 percent of the population is obese — that is, they are 20 percent or more above their maximum desirable body weight. The incidence of obesity has doubled in the past decade, and much worse is to come. A recent report predicted that by 2005 one-quarter of British women and almost one-fifth of men will be obese.

In much of Europe, obesity afflicts 15 to 20 percent of the middle-aged population. This picture is better for Scandinavia and the Netherlands, where the figure is around 10 percent, but worse for eastern Europe, where it can soar to 50 percent among women. The United Kingdom, France, and Germany each has between five million and ten million inhabitants who are obese.

The possible consequences of adult obesity include diabetes, high blood pressure, high blood cholesterol, coronary heart disease, sleep apnea (a life-threatening problem in which breathing stops), gallbladder disease, chronic heartburn, arthritis, certain cancers, and depression. Socially, fat people can be as successful as anyone else — think of Santa, for example — but their shape can undermine their self-esteem, particularly in western societies, where thin is fashionable.

Doctors have declared war on obesity. Nowhere is the battle of the bulge waged more seriously than in the United States, where obesity causes an estimated three hundred thousand deaths each year and accounts for at least $69 billion in health care costs, lost workdays, and disability. Americans spend another $33 billion each year on weight-reduction products and programs, offered by a largely ineffective slimming industry. The lack of progress in addressing this problem has provided a powerful spur for efforts to find out why we are getting so fat — in the process shedding new light on Santa's stomach.

Fat Science

Thousands of years ago, fat meant the difference between life and death: being able to store large quantities of energy-dense fuel in the form of adipose (fatty) tissue enabled our ancestors to survive when food was scarce. In the West today, an abundance of food and a sedentary lifestyle mean that variations in the way individuals balance energy intake and output make some more at risk of obesity than others.

When more energy is taken in than is burned by the body during exercise and while maintaining its basal metabolism, the energy required just to keep it running, the excess is stored as fat. If this continues, obesity will be the eventual result. Studying how we crave, store, and use food energy is the key to finding an effective treatment for obesity and perhaps solving the mystery of Santa's paunch. Whether you look forward to your Christmas lunch, dread its effects on the waistline, or both depends on a complicated web of chemical events, from satiety mechanisms at work within the brain to the molecular machinery that lays down deposits of fat in cells.

The conventional view is that whether Santa, or indeed anyone else, feels hunger pangs depends on a range of signals, from the sight of roast turkey and its wonderful aroma to savoring the first taste. Most of all, appetite depends on how the body interprets information from the gut and hormones in the blood to tune our eating habits to energy demands. Since the turn of the century, when it was found that damage to a structure in the brain called the hypothalamus results in obesity, scientists have realized that these urges result from something going on in our heads. Santa, like the rest of us, is a victim of a Stone Age brain and its appetites, as we saw in the last chapter.

Evolution has spent more time optimizing our appetites for the food available on the prehistoric African savanna than for today's fast-food culture. As a consequence, appetites evolved to supply the needs of the body in a tough environment where fat and salt were relatively scarce and famine never far away. We have inherited a Stone Age appetite, and as a consequence it is easier to overeat fats such as ice cream than carbohydrates such as potatoes, and much easier to overdo carbohydrates than proteins. Whereas the protein requirements of primitive humans ranged from 14 to 20 percent of daily intake, the fat appetite was left largely unregulated, because its immediate benefits far outweighed long-term hazards such as heart disease, which have now become significant as more people reach old age.

Aside from the mechanisms that control the appetite itself, there are others at work that process and store fat. The first step toward laying down another roll on an expanding midriff takes place in the stomach, where the enzyme pepsin acts on food to break down proteins. When the acidic contents of the stomach enter the small intestine, bile and enzymes are released. They emulsify fats, or triglycerides, as biochemists call them. Sugars and protein are similarly broken down, and together these chemical fragments pass into the bloodstream, where they are distributed around the body. Obesity develops when appetite outruns the body's nutritional needs and excess fat accumulates in specialized fat storage cells called adipocytes.

A locally acting signaling chemical called a prostaglandin tells cells when to store fat, according to the studies of a Salk Institute/Harvard Medical School research team led by Ronald Evans. Prostaglandin sends this message to deposit fat by docking with a receptor, a specialized protein found within the nucleus of a cell. This union acts like a switch, turning on genes located in the nucleus and instructing the cell to develop into a fat cell. An enzyme called lipase is then secreted into the bloodstream by the fat cells, breaking down the circulating emulsion of fat (triglycerides) into its basic chemical building blocks, fatty acids and glycerol. In this form, they can be taken up by the cells and then reconverted into triglycerides to create adipose tissue. A second type of lipase present in the cell can be summoned by the body to break down the stored fat and convert it back into energy.

Recently another part of the fat puzzle was fitted into place. A second receptor, found in the liver, appears to function as a sentinel that monitors fat consumption. Eat a slice of pizza, and the sentinel's alarm goes off. The body responds by triggering a fat-burning process.

This discovery casts new light on the effects of fat. It turns out that the sentinel is not able to sense all fats, and thus some go undetected. Interestingly, polyunsaturated fats are great triggers, and therefore might be considered "good fats" because they promote their own removal, Evans explains. By contrast, saturated fats are poor triggers, and therefore could be considered "bad fats." Thus what Santa eats is just as important as what he doesn't eat.

Of all the astonishing advances in understanding why some of us lay down greater fat deposits than others after a Christmas blowout, the most impressive have come from the discovery of genes linked to obesity. When there are defects in these genes, cravings for fatty foods such as ice cream can go on long after the body's capacity to use their calories has been exhausted. With each and every one of these molecular clues comes a new target for drug designers seeking the discovery of an antifat pill — and another hint to why Santa is so plump.

Fat Mice

Our understanding of obesity owes a great deal to strains of fat mice and efforts by scientists to track down the specific genes that make them overeat. You may ask what pudgy rodents have to do with humans, let alone Santa. First, they breed quickly, and you can do things to them you would never dream of doing to a human, let alone Santa. Second, it turns out that if a gene does something useful for mice, chances are one very similar is also at work in humans, a phenomenon scientists explain by saying that nature tends to be highly "conservative."

This conservatism stands to reason, since we all evolved from common ancestors, which in turn evolved from a primitive bug that lived some 3.85 billion years ago. If a piece of molecular machinery that controls cell division works for yeast, for example, why bother to change it for humans? In other words, nature obeys the maxim "If it ain't broke, don't fix it."

To date, a handful of genes that influence obesity have been discovered in mice. Known as tubby, fatty, diabetes, obesity, or agouti, each gene represents a different molecular mechanism for maintaining the right amount of stored energy, in the form of fat.

The race to uncover the genetic basis for obesity can be traced back to 1950, when a team at the Jackson Laboratory in Bar Harbor, Maine, discovered a fat strain of rodent they dubbed the *obese* (*ob*) mouse. It was not until 1994, however, that the reason why these rodents were so fat became clear — it was due to a defect in a single gene, also named ob. The discovery was made by six scientists from the Howard Hughes Medical Institute and Rockefeller University in New York.

The cover of the journal *Nature* that announced the advance made by Jeffrey Friedman, Stephen Burley, and their colleagues showed a scale on which two slim mice were outweighed by one grotesquely fat colleague. This mouse looked like a furry ball because he suffered from a fault in the ob gene, the blueprint for a protein called leptin. The protein is a hormone that tells the brain how much fat is deposited in the body and thus is part of the "stop eating" message.

The researchers measured the amounts of leptin in the blood and found that fat mice with a defective ob gene do not make the hormone. Further evidence of leptin's role came when Friedman and his team found leptin in the blood of six lean humans. "Our findings indicate that when ob is defective, leptin is not made and does not transmit its signal to stop eating," says Jeffrey Halaas of Rockefeller University. Injections of the protein cut the body weight of these mice by 30 percent after two weeks of treatment.

Friedman likened the body's fat-control mechanism to a thermostat: the "fatstat" senses leptin levels, which act as a chemical barometer of the amount of body fat, and adjusts fat deposits accordingly, by regulating how much food is eaten and how much is turned into energy in the body.

Meanwhile, other work has shown that our hunger is stimulated by the action of another molecule in the brain called neuropeptide Y (NPY), which acts on feeding receptors in the hypothalamus. Subsequent research showed that when NPY was genetically removed from the obese mice, the mice quickly slimmed down.

Although leptin helps the brain assess the size of fat deposits in the body and NPY stimulates appetite, these are but two players in what Friedman has called the alphabet of weight control. For example, although high levels of NPY can stimulate feeding, the molecule is probably not involved in normal feeding. It acts by docking with a receptor called Y5, and when that receptor is knocked out in a mouse, the rodent retains a normal appetite, suggesting that NPY is not critical in telling animals to eat.

Although we have yet to understand the details of how this alphabet spells out FAT, the original discovery of the ob gene and leptin caused a sensation. Injections of leptin can fool the body into thinking it has too much fat, reducing appetite and increasing energy consumption. The biotechnology company Amgen reportedly agreed to pay a $13 million signing fee to Rockefeller for rights to the gene.

Since the pioneering work at Rockefeller University, the human equivalent of ob mice have been reported by Sadaf Farooqi and Stephen O'Rahilly of Addenbrooke's Hospital in Cambridge, England. Cousins ages two and eight, born in Britain of Pakistani parents, provided the first proof that leptin plays a crucial role in weight control, an important landmark in obesity research.

The eight-year-old weighed 14 stone (196 pounds) and had already had liposuction on her legs to help her to move around. The two-year-old weighed 4½ stone (63 pounds), and more than half his body weight was fat. From the time they were babies, both children ate nonstop as a result of low levels of leptin. This discovery not only pointed to an obvious treatment — once-daily leptin injections — but also raised the possibility that even people who are mildly obese may carry variants of this genetic defect.

The other implication of these first human cases is that Santa could indeed suffer from a defective ob gene, according to Friedman's colleague Stephen Burley. In other words, Santa's fatstat is deprived of valuable information about the deposits carried around his waist. "Santa is also a diabetic," Burley says. His pendulous tummy would lead to adult-onset diabetes, the most common type, because excessive fat has been found to promote resistance to the insulin hormone that regulates blood sugar levels. The pancreas begins to make more and more to keep up but eventually runs out of steam, so there is not enough insulin. The result is diabetes, when sugar metabolism goes awry.

Ob is not the only defective gene that could be responsible for Santa's girth. The mechanism of action of another, called agouti, was reported in January 1997 by a team at Oregon Health Sciences University. Faults in this particular gene result in a form of obesity in mice, caused by overeating, that has a greater similarity to common forms of human obesity, in that animals may become 20 to 50 percent heavier than normal. "The mice develop a moderate form of obesity rather than the morbid kind of obesity induced by some of the other mouse genes associated with obesity that have been discovered to date," says Roger Cone, senior author of the study. "Most humans who are overweight are moderately heavy, not morbidly so."

The agouti gene was so named because it has long been known to cause a yellow coat color in mice and other animals. The gene does this by interfering with a key molecular docking site, or receptor, on the surface of pigment cells in hair follicles. This site, called the MSH receptor, is responsible for stimulating the production of dark pigment, called melanin. The protein made by the agouti gene blocks the receptor and thus the production of melanin, which allows yellow pigments (phaeomelanin) that are already present to be readily seen.

Now it appears that the protein made by the agouti gene, a mutated version of a naturally occurring gene, also blocks an MSH receptor in brain cells, interfering with a signaling pathway in the hypothalamus, a region of the brain involved in feeding. As Cone explains, "When mutated forms of the agouti gene are inappropriately expressed in the brain, normal feeding behavior is disrupted, and the mice become obese and develop early symptoms of diabetes."

The animal overeats because the mutant agouti gene interferes with a signaling pathway involving melanocortin, a neuropeptide, which is a small stretch of protein that carries information from one neuron to the next in the brain. In a normal mouse, when melanocortin binds to a particular kind of MSH receptor, called the MC4 receptor, it exerts an inhibitory effect on eating and causes the animal to stop feeding.

If Santa had a mutated agouti gene, he would fail to receive an inhibitory signal from the brain and thus consume excessive amounts of seasonal fodder. This hypothesis has important support from a study conducted by Millennium Pharmaceuticals with Cone's team: specially bred mice lacking in the MC4 receptor overeat just like the agouti mutants.

Another candidate gene in the alphabet soup of pathways that influence feeding was found by researchers at the University of California at Davis Medical Center, Duke University Medical Center, and Centre National de la Recherche Scientifique near Paris. As Duke's Richard Surwit explains, "This is a gene that may determine whether a high-fat diet makes you fat or not. It may explain why certain people can eat whatever they want and never get fat, while other people can't. We believe this is at the heart of what happens in people who do get fat."

Given the somewhat dull name of uncoupling protein 2 (UCP2), the gene contains the blueprint for a previously unknown protein that exerts an effect on the energy expenditure of the body. This expenditure takes three forms:

supporting physical activity, maintaining resting metabolic rate, and producing heat, or thermogenesis. UCP2 apparently affects the last of these by burning excess calories in the diet as surplus body heat before those calories can be stored as fat. As a result, people who have more of the protein burn more calories, while people who have less of it store calories as fat.

Even small changes in heat production could potentially have a big impact on a person's weight over the long term. For a typical adult, just a 1 percent reduction in body heat could translate into a five-pound weight gain over the course of a year, all other things being equal. By the same token, raising body temperature by just 1 percent would result in a five-pound weight loss. Obesity could be caused by an alteration in body temperature of one-tenth of one degree. The scientists think of the UCP2 gene as a missing link that helps us understand the connection between body temperature and weight.

Although the discovery isn't a cure-all, it could lead to simple therapies for some people who are overweight. These therapies would increase their expression — that is, the use in the body — of the gene, so that they would burn off more of the calories they eat as body heat. It is even theoretically possible to increase expression of the gene only in targeted tissues to reduce the size of fat deposits in certain areas of the body — an idea Surwit calls "genetic body shaping."

Reshaping Santa

Remodeling Santa's body does sound rather excessive, but scientists have already demonstrated in mice that it is possible, through genetic engineering, to produce a creature that can eat huge amounts of fatty food yet never get fat. Stanley McKnight's team at the University of Washington School of Medicine in Seattle bred mice lacking a single gene that is involved in the complicated and delicate balance of metabolism. This gene, called RII beta, codes for a segment of an enzyme, called protein kinase (PKA), that regulates fat storage and metabolism.

Knocking out the gene makes the mice more sensitive to hormones that break down white fat, the kind that packs on pounds and is responsible for middle-age spread. But more important, it affects a process that takes place in a related kind of tissue, known as brown fat, McKnight says. Brown fat is the body's generator, converting fat into heat. (It is called brown fat because its cells have large numbers of mitochondria, the "power packs" of cells, which tint the tissues brown.) In mice who have lost the RII beta gene, PKA is overactive, stimulating fat all over the body to burn off energy. The brown fat kicks into high gear, using stored white fat for fuel. "Brown fat is acting like a little furnace," McKnight says. The result is perpetually slender animals.

These mutant mice were found to store about 50 percent less fat than their normal counterparts when fed the typical low-fat laboratory food. When eating fatty lab fare, the mouse food equivalent of french fries, hamburgers, and premium ice cream, the difference becomes even more pronounced.

The mutant mice have about half the fat of normal mice, although they do not have fewer fat cells. They also have a higher metabolic rate and a slightly higher body temperature. "The more food and calories they take in, the more obvious it is that they won't get obese," McKnight observes. "They're protected from obesity."

Other labs have developed genetically skinny mice, but most of these mice have had physical problems related to the genes that were knocked out. McKnight's mutants, however, appear to be normal, living as long as other mice and maintaining their fertility. Using genetic engineering, it may even be possible to interfere with PKA to ensure that we are all fast burners, those rare and infuriating people who can eat all the fried foods they want and never gain weight. It is fascinating to consider whether the popular image of Santa will then evolve into a more svelte Kriss Kringle, along with the rest of a thinning population.

Why Santa Laughs in the Face of Old Age

Two other aspects of Santa deserve scientific comment. First, his jolly nature. That merriment could be another result of his genetic propensity to obesity. Plump, bouncing baby boys are more likely to grow into happy men, while skinny babies often get the blues in later life, according to a study by Ian Rodin of Southampton University, England. Rodin studied men born in Hertfordshire, England, between 1911 and 1930. He investigated how factors during pregnancy and the first years of a baby's life could alter brain chemistry and hormonal responses that could determine whether an individual has a sunny or gloomy disposition later in life. "There is no doubt that depressive disorder is associated with low birth weight in men, and the heavier the baby the fewer the depressive episodes," he reported.

Another puzzling thing about Santa is the fact that, though elderly, he never ages. He seems to have found a way to arrest the aging process so that, as the years roll past, he remains sprightly enough to clamber down millions of chimneys.

One school of scientific thought says that life span is determined by a fixed amount of metabolic activity: eat less, slow your metabolism, and you may live longer. Experiments on rodents and monkeys provide powerful support for this idea. However, Santa's expansive waistline rules out this particular explanation.

Age at death is at least partly under genetic control, and the race is on to identify the genes and molecular mechanisms that are responsible. Scientists and pharmaceutical companies need look no further than Santa. There is good evidence that he may have beaten them to the prize of a Methuselah pill.

Enough is now known that we can speculate about how Santa stays so old but grows no older. One idea is that he has developed a method to manipulate structures called telomeres, which are found on the ends of chromosomes and, rather like the plastic bits at the ends of shoelaces, stop them from "fraying." Telomeres shorten with repeated cell division until, when they have dwindled to a certain point, the cell can no longer divide and becomes senescent.

Santa may have found a way to prevent telomeres from shortening, perhaps by activating an enzyme called telomerase. The only problem with this theory is that even the cells of the oldest individuals still have quite a few more cell divisions to go, suggesting that there is more to the antiaging story than this.

Santa also may have spent a great deal of time studying *Caenorhabditis elegans,* a little worm called a nematode. Scientists have discovered in nematodes genes that control the biological timetable from the cradle to the grave. The first ever life-extension mutation, in a gene called age-1, was found by Thomas Johnson at the University of Colorado in Boulder. Others have since been uncovered, including daf-2.

Work by the "worm group" at the University of San Francisco has shown that age-1 and daf-2 mutations act in the same life span pathway and extend life span by triggering similar if not identical processes. An important clue to what is going on in these aged worms came from the work of Gary Ruvkun at Massachusetts General Hospital, who believes that daf-2 may regulate glucose (sugar) metabolism.

According to Ruvkun, when daf-2 is defective, the worms are unable to respond to "worm insulin" and enter a state of hibernation, which is usually triggered by starvation. However, just a slight defect in the gene means that they do not hibernate but instead shift their metabolism so that they can live longer. Ruvkun's research suggests that how fast humans age may be intimately tied to how we burn the calories we eat. Perhaps Santa has found a way to fiddle with his cellular glucose machinery to slow his aging.

A number of potential pitfalls face any wanna-be gene tinkerer in this area. Although some of the long-lived worms seem to retain their first flush of youth, other gene mutations can cause the nematodes to live in slow motion. Using one such mutation, Siegfried Hekimi and his colleagues at McGill University in Montreal created nematodes that live up to 50 percent longer because of a leisurely metabolism.

This particular aging gene does not seem to correspond to what is known about our fat friend. Santa has a frantic workload on Christmas Eve that would have discouraged him from fiddling with the slo mo gene. Nor have I ever seen a claim that he has "hoooo, hoooo, hooooed" rather than "ho ho hoed."

Anyone would be forgiven for thinking that aging researchers have an unhealthy obsession with worms. Worms are, in fact, much quicker and easier to work with than humans, though progress is being made in understanding the genetics of human aging as well. In 1996, scientists found the first human gene known to affect the aging process — one responsible for a rare disease called Werner's syndrome.

Sufferers appear to age in adolescence. They turn gray, develop wrinkled skin, lose their hair, and become susceptible to diseases associated with old age, such as cataracts, heart disease, diabetes, and cancer, usually dying before the age of fifty. Darwin Molecular Corporation, a Seattle biotechnology company, and Gerard Schellenberg's team at the Seattle Veterans Administration Medical Research Center showed that these individuals carry a faulty version of a gene for a type of enzyme called a helicase.

Helicases split apart, or unwind, the two strands of the DNA double helix, a process that must take place in healthy dividing cells if they are to pass on their genetic material, in the form of chromosomes, to daughter cells. If a lack of helicase accelerates aging, Santa may have found that by introducing extra copies of the gene into his body, he can live longer. By using genetic engineering to stay a sprightly sixty-something, Mr. Claus can ensure that he will be able to deliver presents for many years to come.

7

THE BETHLEHEM STAR

Silent night! Holy night!
Guiding star, lend thy light!
J. Moier

The Bethlehem star was probably not the brilliant object portrayed on Christmas cards. It seems that King Herod and all his "chief priests and scribes" did not notice it. St. Matthew used no adjectives such as "bright" to describe it in his Gospel. Only in the early, less reliable Christian literature does the star start to dazzle.

Many and various ideas have been put forward to explain the heavenly apparition that heralded the birth of Christ: a comet, a star birth, a star death, a conjunction of planets, an apparent hesitation in a planetary orbit, even a sighting of the then-unknown planet Uranus. Two thousand years after it was first seen by the Wise Men, astronomers are still in hot pursuit of this heavenly body.

When gazing up at the night sky, it is easy to understand why some people have believed their destiny to be in the grip of the heavens. Two millennia ago, this was felt much more keenly, and the Magi took an interest in the cosmos that is quite alien to the thinking of today's astronomers. We must journey back in time to this culture and, like good anthropologists, attempt to see the heavens through ancient eyes and minds to understand why the star was of significance to the Magi's Babylonian society. First we have to assemble the meager clues to the identity of the star that appear in the Bible. The Gospel according to St. Matthew (2:1–12), the only Gospel that mentions the star, states that "in the time of King Herod, after Jesus was born in Bethlehem of Judaea, Wise Men from the East came to Jerusalem, asking 'Where is the child who has been born king of the Jews? For we observed his star at its rising, and have come to pay him homage.'"

Some theologians dismiss this reference to the star as simply a story made up to satisfy the Old Testament prophecy that "a star shall come forth out of Jacob and a scepter shall rise out of Israel." Fulfilment of such a prediction would have provided succor for the faith. It certainly would have merited a mention, the biblical equivalent of "I told you so!" Matthew's Gospel is full of references to the Old Testament, yet there is no such "fulfilment statement" regarding the star. If we conclude, then, that this heavenly apparition was real, rather than something cooked up to satisfy an Old Testament prediction, what did the Wise Men see? To identify the star of Bethlehem, all we have to do is assemble the clues in the Bible, work out what the heavens looked like on the appropriate night, and then search for star candidates. Easy.

If only it really were that simple, explains Colin Humphreys of Cambridge University. "The difficulties of treating the star of Bethlehem as a real astronomical object should not be underestimated," he says. According to another Bethlehem star theorist, David Hughes, reader in astronomy at Sheffield University, England, to make any progress you have to weigh the evidence and be selective: "When you come to interpreting the facts, you have to pick and choose. You can't take them all on board."

One example of what David Hughes means is the description of the star's movements in Matthew 2:9–10: "There, ahead of them, went the star that they had seen at its rising, until it stopped over the place where the child was." This kind of evidence stirs dissent. According to Hughes: "Astronomical objects are a huge distance away and do not wander in front of people and then stand over and point out a specific house in a small village like Bethlehem." Humphreys, however, believes this biblical reference hints at the star's real identity: "I spent quite a long time scouring ancient historical and astronomical literature and found two other references in which a star was said to 'stand over' a place, both of which are comets." Comets are chunks of frozen matter that sweep through elongated orbits around the sun, warming to form a luminous tail of charged particles on their inward-bound trip.

Perhaps the star never existed. Did Matthew invent it to embellish the Nativity story? If so, he certainly did not describe it in the same exaggerated way as James, whom tradition describes as the brother of Jesus, and who was the author of an Apocrypha gospel. A similar reference is found in the letters of Ignatius, the first-century bishop of Antioch, who writes of a star that "outshone all the celestial lights, and to which the Sun and Moon did obeisance." I am inclined to believe, as Hughes put it, that "Matthew was a straight guy, telling it like it was."

When Was Jesus Born?

Identifying the star would be easier if we knew exactly when Jesus was born. Then we could use a computer program to extrapolate from what we can see of the heavens today to what the Wise Men saw of them on that historic night. However, we don't have a precise date for Christ's birth.

Strange as it may seem, Jesus was not born in the year A.D. 1, despite the fact that "A.D." stands for anno Domini ("in the year of our Lord"). Our present calendar is a modification of the one introduced by Julius Caesar on January 1, 45 B.C. The Roman system of dating *ab urbe condita* (from the foundation of Rome) dates to the first century B.C. That changed in the sixth century when a monk living in Rome, Dionysius Exiguus, proposed that the Christian era should date from a unique event of far-reaching religious significance, the supposed year of Christ's birth. His system marked the origin of the A.D. sequence we now employ. Unfortunately, the monk overlooked four years of the rule of the emperor Octavian when he tallied up the history of the Roman empire. That suggests Christ was born around four B.C.

To pinpoint the year and date of Jesus' birth, one could alternatively work back from his Crucifixion. According to the Gospels, this took place during the rule of Pontius Pilate as governor of Judaea, which ran between A.D. 26 and 36. (So says the celebrated Jewish historian and pharisee Flavius Josephus.) By dating a blood-red lunar eclipse that biblical and other references suggest followed the Crucifixion, Colin Humphreys determined the date Jesus died on the cross as Friday, April 3, A.D. 33. But counting back from this point presents a problem in that we don't know with sufficient accuracy how old Jesus was when he died. Luke says he was "about thirty" when he started his ministry. In another biblical reference, Jesus is told, "You are not yet fifty."

The Bible provides other clues and secondary evidence to date the first Christmas. Jesus was born during the reign of Augustus Caesar, which narrows our search down (though not much) to sometime between 44 B.C. and A.D. 14. Another clue is given by Matthew and Luke, who agree that Jesus was born during the rein of King Herod. It is generally accepted that Herod the Great died in the spring of 4 B.C., though other dates have been put forward (5 B.C., 1 B.C., and A.D. 1). He was replaced by his son Herod Antipas (21 B.C.–A.D. 39), who ruled throughout the ministry of Jesus. That also shrinks the timespan within which the nativity took place but nowhere near enough to reveal what exactly the Wise Men were chasing.

Matthew 2:16 provides another hint for the star detectives to mull over: "Herod . . . killed all the children in and around Bethlehem who were two years old or under, according to the time that he had learned from the Wise Men." Thus Jesus was probably born at least two years before the end of Herod's reign.

Another point on Christ's time line comes from a reference to a Roman census in Luke 2:1–7 that induced Joseph and Mary (who was "great with child") to travel to Bethlehem: "In those days a decree went out from Emperor Augustus that all the world should be registered. This was the first registration, taken while Quirinius was governor of Syria."

Here we run into some difficulty. There is no official record of a census by Publius Sulpicius Quirinius, who became governor in A.D. 6 (by which time Herod the Great was dead). He did conduct a census, but that was of Judaea, not of Galilee, in A.D. 6–7. Luke 2:1–5 refers to a census by Emperor Augustus Caesar around the time of the birth, and indeed three well-documented censuses were conducted for Augustus in 28 B.C., 8 B.C., and A.D. 14, but these were apparently only for Roman citizens. These seem wide of the mark, but Humphreys points out that there is a reference to a census of allegiance to Augustus at the time of the birth of Christ. This was recorded by fifth-century historian Orosius, and Josephus also appears to mention the same event. Perhaps it was this tally to which Luke was referring.

Confused? A great deal has been written about precisely what census is documented in the New Testament. There is no need to go into any more detail here save to say, yet again, that this muddle underlines one of the problems facing anyone who wants to identify the real Bethlehem star: what we know is only as reliable as the few surviving sources. Indeed, even if we accept that the Apostle Matthew composed the first gospel, research by psychologists has shown repeatedly that eyewitness testimony can be subject to error.

The Bethlehem star papers that appear in scientific journals each Christmas only serve to emphasize that the birth date is still very much open to debate. However, it is reasonable to conclude that, using the kind of sleuthing described above, Jesus was born sometime between 4 and 7 B.C.

This window rules out some candidate "stars," such as Halley's comet in 12 B.C. or the conjunctions of Venus and Jupiter on August 12, 3 B.C., and June 17, 2 B.C. The latter is a pity, because on May 17, 2000, Venus will seem to merge with Jupiter again, marking a possible second coming — though it will be too close to the sun to see with the naked eye.

We can use similar detective work to narrow down the

time of year that Jesus was born. December 25? Not likely. A birth date at some point in the spring or even autumn seems a better bet, in the light of another icon of the Nativity, as referred to by Luke: "There were in the same country shepherds abiding in the field, keeping watch over their flocks by night."

Shepherds were most likely to be with their flocks during the lambing season in the spring or when the flocks were being collected in the autumn. Indeed, some Christians have already celebrated April 17, 1995, as Jesus' two thousandth birthday. Others, such as David Hughes, stick to autumn "because the birth of John the Baptist was thought to have been in late March, and Jesus, his cousin, was six months younger."

Bethlehem Stars

Armed with these parameters in which the birth of Jesus probably took place — sometime around September or March between 4 and 7 B.C. — we can draw up a list of candidates for the Bethlehem star.

A comet was suggested as long ago as A.D. 248 by Origen, the celebrated Christian writer, teacher, and theologian. Perhaps it was the "broom star" (*sui-hsing*) — so called because the comet's tail looked as though it was sweeping the sky — that was noted in 5 B.C. by Chinese astronomers and was recorded in the official history of the Han dynasty. If we accept this reasoning, the first Christmas was in the spring of 5 B.C.

The comet of 5 B.C. appeared in Ch'ien-niu, which, according to ancient Chinese star maps, is the area of the sky that includes the constellation Capricorn. In March or April, Capricorn rose above the eastern horizon when viewed from Arabia and thereabouts.

The Magi had the know-how and background to see the significance of the comet and be motivated to chase it, Humphreys argues. In classical literature, the Magi, important members of the Babylonian royal court in Mesopotamia, are depicted as a religious group skilled in the observation of the heavens. From the fourth century B.C., Babylon had been the center of astronomy. Moreover, it contained a thriving Jewish colony from the time of the Exile in the thirteenth century B.C. onward, so that prophecies of a savior-king, the Messiah, may have been familiar to the Magi.

In the Hellenistic age (322–30 B.C.), some of the Magi left Babylon for neighboring countries, and by the time of the birth of Christ, they lived mainly in Persia, Mesopotamia, and Arabia (now Iraq, Iran, and Saudi Arabia). Humphreys suggests that the Magi who saw the star of Bethlehem probably came from Arabia or Mesopotamia.

Humphreys argues that comets were associated with great kings, and the Magi were known to have visited illustrious monarchs in other countries. Here we encounter a problem with the comet theory. Ptolemy, the second-century astronomer/astrologer from Alexandria, endowed the appearance of comets with a grim significance. His *Tetrabiblos* (the "bible" of astrologers) 2.9 warns, "Through the parts of the zodiac in which their [comets'] heads appear and through the directions in which the shapes of their tails point, the regions upon which the misfortunes impend."

The comet of 5 B.C. was visible for seventy days, which Humphreys says is consistent with what is known of the Magi's journey: the distance from Babylon to Jerusalem was at most nine hundred miles (traveling via a region called the Fertile Crescent), which would have taken between ten and twenty days (a fully loaded camel can handle between fifty and one hundred miles each day).

In Jerusalem the Wise Men would have discussed the significance of the heavenly portents with Herod, says Humphreys. "The story started in May of 7 B.C., when Jupiter and Saturn came together against a backdrop of Pisces, signifying that a Son of God would be born in Israel. They would have told him that this happened twice more in 7 B.C. to reinforce the message, and then they would have told of another event, a triple massing of planets in 6 B.C. and the comet in 5 B.C., which conveyed the message that the birth was about to happen. Hence they jumped on their camels and came to Jerusalem." Humphreys argues that the Magi would have had sufficient time to see the comet in the east, and, by the time they set out for Bethlehem after visiting King Herod in Jerusalem, they could have seen it in front of them as they headed south.

How did the comet direct the Magi where to go? Given the model of the heavens that then prevailed, comets were believed to be below the "heavenly spheres" containing the stars, planets, and so on. Humphreys explains how the Magi might have thought of the comet as hanging stationary over a given spot, particularly if it was low in the sky and its tail was oriented vertically. This interpretation vividly fits Matthew's account: "Lo, the star, which they had seen in the east, went before them, till it came and stood over where the young child was."

There has been some debate over whether the Chinese records imply movement typical of a comet. According to Humphreys, the phrase "a sui-hsing appeared at" implies definite motion. Others have taken the translation literally and suggested that the Chinese saw a point of light wink on in the sky. The latter idea led some British astronomers to suggest that the Chinese mistakenly categorized the object as a broom star when it was in fact a guest star — the thermonuclear flash of a nova. (*Nova stella* is Latin for "new star.") This theory dates back centuries, perhaps even to a hint in *De Vero Anno*, written in 1614 by the great astronomer Johannes Kepler. A few such novas appear each year, when a faint, usually unseen star brightens by a factor of ten thousand or even one million. These outbursts are thought to occur in a binary, or pair of stars, when gases from the larger member fall into the smaller member, triggering a nuclear conflagration. The new star would have appeared in the east several hours before sunrise (Matthew 2:2: "We observed his star at its rising").

However, for the same reasons the comet seems attractive, the nova does not. Matthew 2:9 suggests that the object was later visible in the south when the Wise Men headed south toward Bethlehem after their visit with King Herod. A nova would not have moved that much. The location of the nova also is unlikely, given that the Bethlehem star appeared well away from the disklike plane of our galaxy. The latter is lush with stars — its hazy cross section is seen in the sky as the Milky Way — and likelier to be a stellar nursery.

Another difficulty with both these theories pointing to a 5 B.C. birth date is that the astrologers of the Middle East were preoccupied with the planets, the sun, and the moon. They had little time for novas and comets. At least this is the case if you accept the line of reasoning of some star theorists.

But we have far from exhausted the possibilities. A star death, or supernova, is a suitably dramatic candidate for the

Bethlehem star, one with the potential to light up the night sky. Indeed, in A.D. 1054 Chinese astronomers observed a supernova that was bright enough to see in the daytime. However, this suggestion has been discounted, since the remnant of such a cataclysm, close enough to Earth for it to have appeared bright, would have left a spectacular aftermath, a splash of radio and X-ray wavelengths that would still be visible to astronomers.

Perhaps the Wise Men were struck by a moment of hesitation on the part of Jupiter. This idea, put forward by the late British polymath Ivor Bulmer-Thomas, proposes that the Bethlehem star was Jupiter passing through a stationary point in its trek across the sky. When a planet undergoes such a retrograde motion, it makes a loop against the stars.

This occurs because of the relative position of the planets around the sun. Those planets that are farther out than Earth complete an orbit around the sun more slowly than Earth. Because of this, Earth catches up and overtakes them as it circles in its own orbit, making them appear to travel backward for a while before they revert to their usual progression. Jupiter appears to be stationary at each end of the loop for about a week.

Hundreds of cuneiform texts excavated in Babylon pay tribute to how Babylonian astronomers took a keen interest in such retrograde motions. The natural explanation for St. Matthew's "star," argued Bulmer-Thomas, is that the Magi in their journey from the east to Jerusalem and Bethlehem were observing the motion of a planet. While they were at Bethlehem, it reached a stationary point. The Magi were seeking a king of the Jews, because this motion had been executed by Jupiter, the most regal of all the planets.

Bulmer-Thomas went on to theorize that other celestial

events, involving some of the star candidates, alerted the Wise Men to the stationary point. Three conjunctions of Jupiter and Saturn in 7 B.C. had great meaning at the time. A clay tablet, the star almanac of Sippar, which has been found about thirty miles north of Babylon, refers in detail to the triple conjunction. Then there was the grouping of Mars, Jupiter, and Saturn in Pisces in 6 B.C. and the comet of March or April in 5 B.C.

Forewarned by these celestial events, the Magi may have followed Jupiter from the time it emerged from behind the sun in May 5 B.C. Bulmer-Thomas said that they would have seen the regal planet pass through a stationary point four months later, about the time it took for them to complete their journey. "As they approached Bethlehem in the fourth week of September, they could see that Jupiter was near its first stationary point, and this convinced them that the babe they saw lying in a manger was indeed the Messiah." The date also gives sufficient time to Herod to carry out his massacre of children before his death in March 4 B.C.

The Magi As Astrologers

Like many others, I suspect that the objective perspective of a modern astronomer is an inadequate one from which to hunt the Bethlehem star. Instead, the answer to the mystery might better lie in understanding who the Wise Men were and how they interpreted signs in the heavens.

The astronomy of two thousand years ago had no physical perspective and no astrophysics. The idea that planets differed from stars had not occurred to people. Instead, they were concerned with the relative position and motion of these points of light. The way that they interpreted these points of light is highlighted by how one chooses to translate *Magi,* a Greek word. The Authorized Version translates this as "wise men," but the New English Bible opts for "astrologers." Astrology was, after all, widely practiced throughout the Roman world, especially in the regions of the Near East surrounding Judaea. One consequence is that the Magi were more likely than your garden-variety wise men to possess a detailed knowledge of the night skies, so they were unlikely to have been impressed by a routine event such as the appearance of a shooting star. They may, however, have been moved by something in the night skies that would seem unremarkable to a modern astronomer. This is best understood by looking back at the common origin of astronomy and astrology.

Before the seventeenth century, there was not the sharp dichotomy between astrology and astronomy that exists today. The public is still confused, however, about the difference between astrologers (who always spout ambiguous rubbish) and astronomers (who sometimes do). At the root of both is the cosmic wisdom of the ancients. A holy man would jealously guard knowledge of the heavens because it conferred the ability to foretell the future, albeit to a limited extent. Celestial observations guided him through the seasons, revealing when to harvest and when to move herds, and it helped to predict striking events such as a solar eclipse or the flooding of rivers such as the Nile. Recent studies (to which I will return) have even found links between the time of year one is born and what one becomes, but these effects are weak and can be explained without resorting to mumbo jumbo.

Woe betide anyone who confuses astrology and astronomy today. But when the Wise Men gazed at the heavens they could be forgiven for thinking that they could glimpse something of their destiny. Once we accept that the Magi were astrologers, it becomes apparent that they may not have seen a star at all, or indeed a cut-and-dried astronomical object, but an unremarkable cosmic event that nonetheless had remarkable symbolism.

The suspicion that the star has more astrological than astronomical significance is bolstered by evidence that the star was not very obvious. As mentioned earlier, it was sufficiently unremarkable for Herod to be unaware of it until the Magi informed him of its significance. (Those who prefer a straightforward astronomical interpretation would, of course, disagree and point out that there is nothing to say that Herod missed the star.)

There was, after all, a clear difference between the Magi and the Jewish chief priests: Babylonian society was interested in astrology, while in Jewish society astrology was forbidden — at least according to Deuteronomy 4:19 ("Lest thee corrupt yourselves, and make you a graven image. . . . and lest thou lift up thine eyes unto heaven, and, when thou seest the sun, and the moon, and the stars, even all the host of heaven, shouldest be driven to worship them, and serve them").

Christmas Astrology

If we accept that most Bethlehem star suggestions do not take into account the mind-set of the Wise Men, what kind of astrology was practiced in the Near East during the reign of King Herod? Michael Molnar of Rutgers has studied the Greek astrology that was used throughout the Roman world, including Mesopotamia and Babylonia. He has concluded that "Jesus would have been two thousand years old on April 17, 1995."

His candidate for the "star" is an event that took place on April 17, 6 B.C.: a double occultation of Jupiter by the moon, which occurs when our closest neighbor moves in front of the giant planet. Molnar's studies suggest that this event, though of little consequence to a modern astronomer, was "brilliant" in an astrological sense.

Molnar notes that astrological signs appeared on ancient coinage, notably from Antioch, the capital of the Roman province of Syria. On one side of each coin is a bust of Jupiter. On the other, Aries the ram gazes back at a star. Molnar now believes that the coins are related to the annexation of Judaea by the Romans and suggest that the Romans were aware of important astrological portents involving Judaea. He believes that it is very likely that what he calls "the great portent" of April 17, 6 B.C., was very much on their minds: the Romans were looking for proof that a Roman, not a Jew, fulfilled the messianic prophecy. (Augustus Caesar would indeed assume control of Judaea to fulfill the prophecy a dozen years later.)

Aries appears on the coins because it was linked to Judaea in the symbolism of the day: Ptolemy refers to how Judaea was under the spell of the ram. On this point, however, there is dissent. Molnar's theory runs counter to others, which claim that Pisces is the sign of the Jews. Molnar counters that the critics of his ideas are using a Renaissance source, not one from the first century B.C., which is Ptolemy's source for his *Tetrabiblos.* He adds that there are other sources from Roman times that also support Aries as the symbol of Judaea.

Assuming that Aries is indeed the symbol of Judaea, Molnar needed another ingredient to concoct an astrological recipe for the Bethlehem star: something in the heavens to symbolize the birth of a king. "My initial search for a regal 'star' centered on the star of Zeus, namely the planet Jupiter, which invariably played the central role in horoscopes that had regal implications," he says. He found that the regal symbol of Jupiter did indeed feature prominently in the ancient horoscopes of several Roman emperors.

To identify an astrological portent involving Jupiter, he focused on searching for lunar occultations. Examining the likely time frame, he found only two that took place in Aries and thus Judaea: March 20, 6 B.C., and April 17, 6 B.C. "This finding was regarded as an intriguing coincidence, until I later realized that during the second occultation, Jupiter was precisely 'in the east,' an astrological terminology that Matthew uses to describe the Magi's star." The astrological conditions on April 17, 6 B.C., produced truly impressive portents that could only be appreciated from the perspective of ancient Greek astrology. "If we re-create a horoscopic chart," Molnar says, "we find unmistakable indications pointing to the birth of a king of Judaea. I believe that a horoscope of that day was incredibly ominous — truly messianic."

Jupiter continued moving westerly until August 21, when it "stood still" and started to move to the west. During this phase Jupiter was "going before." Jupiter again "stood still" in Aries on December 29. A few weeks later, it exited the sign of Aries. To Molnar it is evident that the translation of the account of Matthew does not perfectly convey the arcane astrological message of the Magi. Here is a paraphrase (one that agrees with Ivor Bulmer-Thomas) of what Molnar believes was most likely said about the Magi's star by astrologers: "And behold the star which they had seen at its morning rising went retrograde, till it became stationary above in the sky [which pointed to] where the child was."

The mystery of the star has been solved. Well, perhaps not. David Hughes, for one, counters that such occultations take place too regularly to be of great astrological significance. He adds that the "hesitations" in the path of Jupiter take place twice a year. ("How often do you want the Magi to go to Jerusalem?" he asks.) Hughes is most struck by a rival idea: a triple conjunction. One between Mars, Jupiter, and Saturn was historically believed to precede the birth of Christ. In 1465 Jakob von Speyer, the court astronomer to Prince Frederic d'Urbino, posed the following problem to the German astronomer Johann Müller (Regiomontanus): "Given that the appearance of Christ is regarded as a consequence of the Grand Conjunction of the three superior planets, find the year of his birth." Müller was unable to answer. In 1604 the German astronomer Johannes Kepler calculated that the massing of Jupiter, Saturn, and Mars occurs every 805 years, suggesting that they coincide with great events in history.

Hughes argues that the Bethlehem conjunction was not of three planets but of Jupiter with Saturn in the constellation of Pisces. The regal aspect comes from Jupiter, king of the gods, while Saturn represents both the principle of justice and the land of Palestine. Disagreeing with Molnar, he reasons that Pisces is the sign of the zodiac that represents the land of Israel. This conjunction, claims Hughes, signifies a potent brew of divinity, kingship, and righteousness against the background of the Jewish people and the Promised Land. "Putting it crudely, that is why the Wise Men went for Jerusalem," he says.

The Magi calculated the details of the triple conjunction well in advance of its appearance. They could have watched the first conjunction from Babylon in May 7 B.C. but delayed traveling until the end of the long, hot summer. On their way to Jerusalem, they witnessed the astrologically important moment when Jupiter and Saturn were rising at the instant of sunset. This was taken as the moment of birth in the phrase translated in most Bibles as "we have seen his star in the east," which Hughes says has a specific meaning, namely, "we have seen his star rising in the east as the sun was setting."

If this explanation is correct, then the only thing miraculous about Hughes's theory is that the Magi noticed the "star" and made the arduous trek to witness, as they said, the appearance of a new king of the Jews. The bad news, or good news if you are allergic to Yuletide cheer, is that this suggests the real Christmas should be celebrated sometime around September, to reflect the events that took place in 7 B.C. Given the spotty evidence, however, the Bethlehem star debate will no doubt rumble on for many years to come.

The Significance of a Birth Date

Those who believe in star signs would, no doubt, be fascinated to know a precise date of the birth of Christ and whether he behaved like a typical Capricorn, Aries, Pisces, or whatever. Intriguingly, there is evidence of a weak relationship between when we are born and how we fare in later life, again underlying the reason any ancient astrologer who was aware of these effects was indeed a wise man.

Science has given some fascinating insights into how the

timing of Jesus' birthday could have influenced the rest of his life, from the functioning of his immune system and risk of heart disease to intellectual and sporting achievement. Although a case could be made for the influence of the moon (think of tides and the menstrual cycle, for example), so far as most people are concerned, there is only one star of influence, and that is the sun.

Earth's relationship with the sun determines the length of the year and the seasons. Depending on which season you are born in, this can have consequences later in life. A study conducted in Gambia by the Medical Research Council's Dunn Nutrition Unit in Cambridge, England, discovered that death rates among villagers who had passed puberty were far higher if they were born in the wet season than if they were born in the dry season.

The divide was startling. By the age of forty, 62 percent of people born during the dry season were still alive, but only 45 percent of those born during the wet season had survived. During the wet season, malnutrition is high because the food from the previous year has been used up and the next batch of food is still growing. This provides the link between birth date and adult well-being.

The quality of nutrition during a child's nine months in the womb appears to affect the development of its immune system. Later in life those whose immune systems did not have the chance to develop to their full capacity appear to suffer far more from infections and to die from them. This research, led by Andrew Prentice, adds a new dimension to discoveries by scientists at Southampton University, England, who have over the years accumulated evidence to link maternal nutrition with degenerative illnesses such as heart disease later in life.

Birth date can also affect sporting prowess. Ad Dudnik of the University of Amsterdam studied the Dutch youth tennis league and found that half of these relatively elite players were born during the first three months of the year. He carried out a follow-up study of birth date and participation in Dutch soccer and found that a disproportionate number of players were born between August and October, the first quarter of the competition year. If Jesus had been born around Christmas, he certainly would have had more chance of participating in the local soccer league than if he had been born between May and July, according to Dudnik's data. The reason for the association has nothing to do with star signs, but how individuals are grouped together by a selection year and when that year starts.

In sports where advanced physical development matters, the youngest children in any age-group are at a disadvantage because they were born late in the selection year. This explains why, for example, almost half the elite swimmers and tennis players in one study were born in the first three months of the year: in tennis and swimming, the selection year starts on January 1.

Unfortunately for those born around Christmas, another study, this time of birth date and success later in life, showed that if you want to be brainy, it helps to be born during the summer. This weak "astrological" prediction comes from research by Takanari Gotoda of Hammersmith Hospital, London. He studied the birth dates of 2,525 graduates from the faculty of medicine at the University of Tokyo, among the most talented at passing high school examinations. He grouped them by month of birth and then divided this number by the number of graduates he would have expected given the monthly birth records in Japan over the same period, from 1947 to 1971. What emerged was a hump-shaped plot with the peak in the summer, suggesting that those born in the summer do better than those born at other times of the year.

Although the reasons for this seasonal effect on intelligence are not clear, you do not have to look beyond the sun for a possible explanation: babies born during the warm summer months are less likely to be wrapped up, suffer illness, or be cooped up indoors. This allows them to be physically more adventurous and enjoy a more stimulating environment, which has been shown by experiments on animals to be important for brain development.

Intelligence is by no means the only characteristic prone to this kind of seasonal effect. Revolutionaries tend to be born around Christmas (to be exact, they have birthdays between October and April), whereas May to September is the season for reactionaries, according to Michael Holmes of Queen Margaret College in Edinburgh. The explanation again rests on Earth's relationship to the sun.

Holmes's study was based on reactions to the ideas of Einstein and Darwin, before the days of central heating, so the winter-born revolutionary would have been restrictively swaddled. Holmes argues that the winter-born proto-revolutionary would have experienced more freedom to explore on his own initiative with the coming of summer. "The summer-born non-revolutionary, on the other hand, would have enjoyed more freedom at first, when less able to use it, but would have been constrained the following winter when ready to extend his explorations."

Although Gotoda's work suggests that those born in the summer are brainier, it seems that they are less adventurous. Whatever the case, Holmes's discovery seems somehow more appropriate for this chapter, given the impact of the birth of Christ and his official December birth date.

Christmas Cards Get It Wrong (Again)

In Bavaria and the Black Forest, children traditionally go from house to house singing carols, while carrying lanterns and a paper star representing the star of Bethlehem. In the light of the previous discussion, they should perhaps discard the star for an icon or object better able to convey the subtle astrological symbolism.

A stellar announcement is only one of many aspects of a Western account of the Nativity scene that turns out, in fact, to be dubious. According to this account, the Holy Family arrived in Bethlehem late at night, where the local inn had no vacancy. The exhausted couple found that there was nowhere to stay but a stable, where Jesus was born.

This popular version of events is flatly disproved by the text of Luke, argues Ken Bailey, director of the Institute for Middle Eastern New Testament Studies in Beirut. The gospels translate the Aramaic speech of Jesus and his disciples into Greek. The careless translation of key words of biblical Greek has led to misinterpretations, particularly with the burden of a modern Western perspective.

The Bible tells us that Mary gave birth to her first son, wrapped him in swaddling clothes, and laid him in a manger. The traditional interpretation of this event, in the West at least, is the following: "Jesus was laid in a manger. Mangers are naturally found in animal stables. Ergo, Jesus was born in a stable."

But this reading ignores the traditional design of peasant

homes in Palestine and Lebanon at the time, Bailey argues. Farm animals shared the same living area as the family, so that the beasts could help warm the house in winter and be kept safe from thieves. Sometimes there was also an adjoining guest room.

The family lived on a raised terrace (Arabic *mastaba*), while their "central heating" — oxen, donkeys, and so on — resided some four feet below on the floor (*Ka'al-bayt*) near the door. The animals' mangers, or troughs for fodder, were found on this floor or at the edge of the terrace. This reinterpretation still supports the idea of Jesus being born in poverty, says Bailey. "That is, he was born in a simple peasant home with the mangers in the family room. He was one of them."

Bailey also questions the reference to the local inn being full. He explains that the word *inn* in our English versions of the Bible is a mistranslation of the word *kataluma,* which has several meanings. It can mean "inn" but also "house" or "guest room." Reading *kataluma* as "inn" creates several problems. First, Luke uses another word, *pandokheion,* for a commercial inn. Second, the only other use of the noun *kataluma* is in Luke 22:11 and Mark 14:14, where "guest room" better suits the context ("And ye shall say unto the good man of the house . . . Where is the guest chamber, where I shall eat the passover with my disciples?"). Third, if Joseph had stayed at an inn, he would have insulted any members of his extended clan who had remained in the village where his family originated and were therefore duty-bound to offer hospitality to kin. Finally, there is great uncertainty over whether Bethlehem even had a commercial inn.

Bailey insists that *kataluma* should be translated "guest room." He draws on cultural and archaeological data to support his contention that Jesus was actually born in the heart of a Palestinian home: "Joseph and Mary arrive in Bethlehem; Joseph finds shelter with a family; the family has a separate guest room, but it is full. The couple is accommodated among the family in acceptable village style. The birth takes place there on the raised terrace of the family home, and the baby is laid out in a manger."

Denis Alexander, editor of the journal *Science & Christian Belief*, finds this rereading convincing: "If [Bailey's] interpretation is correct (and he certainly persuaded me), then the pictures on many millions of Christmas cards (Mary and Joseph being shut out of the inn and so on) are simply mistaken, not to speak of a few million sermons."

The First Christmas Presents

We know that the Wise Men came to mark the birth of the new king with presents of gold, frankincense, and myrrh. Indeed, given that they were astrologers, they probably cast his horoscope as well.

The modern reader may be puzzled by the gift selection of the Magi. The reason they brought gold is obvious but what was the significance of the other two choices? These resins, which have been collected from trees for more than five thousand years, make sweet-smelling perfume and incense but also were reputed to ward off all manner of ills. That made them useful gifts for someone who had just given birth. They also were in short supply, due to the intense demand, which is why they ranked with gold as offerings suitable for the Christ child.

Frankincense is from the French words meaning "pure incense" and has long been valued for the fragrant smoke it produces when burned. It comes from a spiky tree, *Boswellia carteri,* found in the dry areas of northeastern Africa and southern Arabia, such as the arid highlands of Somalia and the Arabian peninsula. The resin is harvested by nomadic tribes, who scrape the bark of the tree and return to collect the "tears" of solidified whitish resin a few months later.

Myrrh is a yellowish red resin collected from the short, thorny tree *Commiphora myrrha.* The tree, which grows across Ethiopia and Kenya, produces the oily, bitter-tasting resin at the bases of its branches. Other species of *Commiphora* also produce resins, such as African myrrh from *C. abyssinica* and the guggulu resin from *C. mukul.* These fragrant resins are all remarkably similar in biochemical terms and are probably produced by the trees to gum up the mouthparts of attacking termites, to aid repair through their antibiotic properties, and to act as a temporary dressing for damaged bark.

The ancient Egyptians used frankincense to treat wounds and in religious rites, when anointing mummified bodies. During Roman times, when cremation was widely practiced, it was customary to burn frankincense in the funeral pyre. According to Pliny the Elder, this was done not just to appease the gods but to disguise the grim odor. Between A.D. 25 and 35, Celsus, a Roman author, recommended frankincense for treating cuts and bleeding. In his medical encyclopedia, the resin rated a mention as a possible antidote to hemlock. In the seventeenth century, Nicholas Culpeper, an herbalist and apothecary in Spitalfields, London, used frankincense to treat stomach ulcers and bruises.

What about myrrh? Early Egyptian myths describe the resin as the "tears of Horus," the god of the sun and moon.

Ancient Egyptians used the resin to protect mummies. Its antibiotic qualities reduced decay and helped to prevent tissue from falling apart, and disguised the stench of the body. The Eighteenth-Dynasty queen Hatshepsut rubbed it into her legs.

Early Sumerian inscriptions describe myrrh treatments for bad teeth and worms, while the Greeks prescribed it for infections of the mouth, teeth, and eyes, as well as for coughs. Both Greeks and Romans thought that myrrh could cure poisoning from snakebites. The use of a little myrrh in "aromatic" wines was found to prolong shelf life. Both resins have been shown by modern studies to have antiseptic, antifungal, and anti-inflammatory properties, and so make valuable dressings. And their oils seem to cause the bronchii of the lungs to dilate, so they may help relieve lung infections and perhaps mild asthma. More striking still, some researchers have found that eating resin or oil from *Commiphora* or guggulu, lowers blood cholesterol levels. Chinese scientists have discovered that *C. myrrha* can reduce the development of arteriosclerosis — hardening of the arteries — in animals.

Recently Piero Dolara and colleagues at the University of Florence began researching myrrh because of the confused references to the properties and uses of the substance in ancient and biblical texts. Mice were given a control or myrrh, then placed on an uncomfortably warm metal plate. Comparison of their subsequent paw-licking times showed that myrrh is a painkiller, with an action similar to that of morphine. Follow-up studies identified the involvement of two types of chemicals, called sesquiterpenes, with analgesic effects: furanoeudesma-1,3-diene and curzarene. Perhaps this explains why *vinum murratum*, "wine mixed with myrrh," was offered to Christ shortly before the Crucifixion. Traditionally, Hebrews mixed myrrh with wine for a condemned man. Jesus was being offered an anesthetic.

Why December?

> *I have often thought, says Sir Roger, it happens very well that Christmas should fall out in the middle of the winter. It is the most dead uncomfortable time of the year, when the poor people would suffer much from their poverty and cold, if they had not good cheer, warm fires and Christmas gambols to support them.*
>
> JOSEPH ADDISON (1672–1719)

How did the birth of Christ become entangled with winter and its paraphernalia of reindeer, Santa, and the like? There are two factors in the explanation. First is the utter confusion over Christ's birth date, which should be clear from the discussion earlier in this chapter. Second is the pragmatic line taken by the early Christians when faced with assimilating pagan rituals into their own.

The origin of the December date lies in festivals in ancient times, when people feared that winter might snuff out the sun god's rule, allowing the powers of darkness to take over. Illuminated by today's scientific knowledge, this primitive dread is understandable: with the exception of strange life-forms that huddle around volcanic cracks in the ocean floor, or bugs that dine on wet basalt, the living economy of the planet relies on harvesting light from the sun.

The traditional time for celebration, December 25, coincides with pagan festivals and follows closely the winter solstice, when the sun reaches its greatest excursion south of the equator. Late December marks an important turning point in the sun's apparent course, after which the daily quota of sunlight grows longer and stronger.

From ancient times, people have lit candles, bonfires, and Yule logs to help nourish the sun god when he was at his weakest and drive winter and its hardships away. The Roman festival of Saturnalia in mid-December is one example. In keeping with its name — *Saturnus* means "plenty" or "bounty" — the celebration involved feasting, gambling, dancing, and singing in honor of Saturn, the Roman god of agriculture. Hats were worn — though not paper ones.

At these festivals, the master served the slave, a ritual that can still be seen in office parties today. Gifts were given, including branches of sacred wood (which evolved into the switches left to punish bad boys and girls in some seasonal traditions). This festival was swiftly followed by the Kalends, a New Year celebration. Already we can see a template forming for the development of the festivities we enjoy today.

The pagan Roman emperor Aurelian proclaimed December 25 as "Natalis Solis Invicti," the festival of the birth of the invincible sun, which was marked by chariot racing and decorations of branches and small trees. In pagan customs dating back to before Roman times, evergreens such as holly and ivy were used for decoration and presents. These plants stay fresh-looking during the winter months, so they were thought to be linked with wood spirits and vitality. As with other primitive holiday trappings, these plants were assimilated into subsequent traditions. The thorns of the holly, for example, were linked with Christ's crown of thorns. Kissing under the mistletoe is another custom that dates back to pagan times, when the plant was associated with fertility. The druids considered mistletoe to be sacred, especially when it grew on oaks, their most revered trees.

People living in ancient Britain and Scandinavia held midwinter sun festivals called "Yule" or "Jol," the origins of the words *Yuletide* and *jolly*. Nor did Germanic tribes of northern Europe miss out on the fun. They assembled in midwinter for feasting, drinking, religious observances, and simply enjoying themselves. White-bearded Odin was thought to roam around punishing evildoers during the celebration in December, prompting attempts to appease him with gifts.

These pagan midwinter festivals remained popular centuries after Christ was born, and early Christians were unwilling to relinquish them. When the church found it impossible through repeated bans to abolish all pagan customs, it "Christianized" a number of them, divesting them of their worst features. That created another problem: since no one knew the actual date of Jesus' birthday, some people celebrated it in the spring and others in the winter.

The first mention of Christmas Day, as far as we know, was in the Roman calendar Chronographus Anni CCCLIV (chronography of the year 354). Within half a century, Christmas Day had become an important date in the Christian year, with December 25 fixed as the "natural" date so as to exorcise the earlier pagan festival of the winter solstice.

And the reason Christmas is abbreviated to Xmas? It started out as an ecclesiastical abbreviation that was used in tables and charts. The first letter of Christ in Greek (*Khristos*) is chi, which is identical to our *X*. Thus *X* stands for Christ, as in Xmas.

8

SNOW

Have you marked but the fall o' the snow
Before the soil hath smutched it?
O so white! O so soft! O so sweet is she!

BEN JOHNSON, CELEBRATION OF CHARIS, IV (HER TRIUMPH)

NO CHRISTMAS WOULD BE COMPLETE WITHOUT it. Even if you live in Los Angeles, you can find tons of the stuff, albeit on cards or the spray-on variety adorning store windows.

Snow also can decorate our language. We call upon it for its utility as a metaphor. It is a symbol of purity, isolation, and transience. Think of Snow White, those ranks of doomed snowmen, or the winter scenes on seasonal cards.

Snow is a commodity we usually remember for the pleasures it offers or the disruption it can cause to traffic and airports. But its contribution as a natural resource is mostly taken for granted. One-quarter of the globe, or 48 million square miles, is snow covered at some time during the year. Worldwide at least one-third of the water used for irrigation comes from snow. In places such as the western United States, for example, the figure is much higher, as much as 75 percent.

Scientists study every aspect of the cycle that gives us snow, from the seeding of a single crystal high in the sky to secrets buried deep in the resulting snowpack, a "time machine" that can be used to investigate climate change over periods of hundreds of thousands of Christmases. The British Meteorological Office hunts for flakes with *Snoopy*, an ancient Lockheed C-130 Hercules with a long, instrumented snout. In Japan researchers are creating snowstorms in the laboratory. In the United States snow studies can help to track pollution or forecast how much irrigation will be available to farmers in the West. Meanwhile, Germany's Fraunhofer Institute for Chemical Technology has developed environmentally friendly forms of artificial snow.

Watching every delicate flake, understanding it, and forecasting it — all this is a major undertaking on which our society depends. Snow is big business, whether you are skiing, trying to keep automobiles moving in a major city, or even gambling on it. One British bookmaker, William Hill, often takes bets on a white Christmas. Winnings that run into thousands of pounds can rest on a light seasonal sprinkling in London. Society is, as a result, now beginning to exert control over snow. Meteorologists seed clouds with silver iodide to induce snowstorms, while ski resorts exploit bacteria to make snow at higher temperatures. And there are even more fantastic schemes afoot to guarantee that traditionally snow-free areas get a white Christmas.

The Seasons

Natural snowfall, "when Frau Holle shakes her featherbed," as some Germans put it, depends on a combination of

Earth's orbit around the sun and a 23.5-degree tilt in its axis. Together, the orbit and tilt conspire to give us our seasons, when there is a change in the direction of the sun's rays relative to the ground and in the number of hours of sunlight each day.

Astronomers talk about equinoxes, when the length of the day and night are equal, and solstices, when the sun is at its highest and lowest in the sky at midday. These events have been used to indicate the start of each season. Thus in the Northern Hemisphere, spring is deemed to start at the vernal equinox (near March 21), summer at the summer solstice (near June 21), autumn at the autumnal equinox (near September 21), and winter at the winter solstice (near December 21).

As Earth sweeps around the sun, the Northern and Southern Hemispheres trade places in receiving the sun's rays most directly. In June Earth's axis is tilted such that the Northern Hemisphere "leans" toward the sun, receiving its rays more directly and for longer periods of the day. This is the summer season in the North. In December the axis is tilted such that the Southern Hemisphere "leans" toward the sun — resulting in summer in the Southern Hemisphere and winter in the Northern Hemisphere.

The differences in the energy received from the sun at any one place on Earth is due to the height of the sun in the sky at midday and the associated length of the hours of sunlight, which are highest at the summer solstice and lowest at the winter solstice. These days do not, however, coincide with the hottest or the coldest weather, respectively. In reality, inertia in Earth's response to sunbathing delays the warming and cooling by several weeks.

Earth has a slightly elliptical — oval — orbit, so it is closer to the sun during the Northern Hemisphere's winter. However, the variation in distance is so small — from 147 million kilometers to 152 million (five million kilometers, or three million miles) — that this seasonal influence is overwhelmed by that of the tilt and other factors. For example, although the Southern Hemisphere's winter occurs when the sun is most distant, the larger mass of water in this hemisphere tends to make winter warmer at equivalent latitudes.

The Stuff of Snow

Snow is made from what is, without doubt, one of the most bizarre, complex, and misunderstood of all substances. We all take it for granted, but for scientists water is a constant source of surprise. It is only by first understanding this ubiquitous liquid that we can approach the science of snow.

Water remains a liquid at room temperature, when substances made of similar-size molecules to those that form water — ammonia, methane, or hydrogen sulfide — are all gases. The boiling point, melting point, and heat-conducting abilities of water also are eccentric, being far higher than for any comparable substance. For example, boiling a pint of water takes far more energy than boiling other liquids.

The misbehavior does not end there. Most substances shrink when you cool them, but ice takes up more space. From the perspective of the passengers of a cruise ship, this is bad news: it means ice floats on water. But when it comes to living things in general, this is very good news. If ice were denser than liquid water, our lakes would freeze from the bottom up. Fortunately for things that swim, paddle, and wallow, an insulating skin of ice forms on a lake to protect the waters below so they remain liquid. Water also has dazzling properties as a solvent. As whiskey drinkers know, it mixes readily with alcohol. Sugar, salt, and other minerals dissolve easily. Water is an ideal medium to transport nutrients into cells. This is why the search for alien life is closely allied with the search for alien water.

All of its strange properties can be understood in terms of water's component molecules and their "social" inclinations. In 1784 the English natural philosopher and chemist Henry Cavendish (1731–1810) described the chemical composition of water, a combination of hydrogen and oxygen, marking the first step toward understanding the molecular structure of water, and thus its properties.

Hydrogen and oxygen are elements, two of the hundred or so basic chemical building blocks of matter. The smallest portion of an element that can exist is an atom, which consists of a central positively charged nucleus, surrounded by one or more diffuse shells of negatively charged electrons that cancel its charge. More than a billion atoms would sit on the full stop at the end of this sentence.

Atoms do not stick together in a haphazard way. Those of one particular element will combine in only certain ways with those of others. At its simplest, a link, or bond, between atoms in a molecule is due to their sharing a pair of electrons, providing a kind of negatively charged "glue" to hold the positively charged nuclei together.

The special properties of water are the result of the way negative electrical charge is smeared over the molecule. We now know that each water molecule consists of one oxygen atom and two relatively tiny hydrogen atoms, arranged in a V. The positively charged nuclei of these three atoms are "glued" together by negatively charged electrons. However, the oxygen is so greedy for the attention of electrons that in the case of water, the two hydrogens are denuded of their negative charge, leaving their positively charged nuclei exposed. That exposed positive charge is strongly attracted to other electrons, notably those in the oxygen atom in an adjacent water molecule. Two pairs of electrons stick out from the oxygen atom like rabbit ears, serving as sticky sites for electron-starved hydrogens on neighboring water molecules. The resulting so-called hydrogen bonds can strap each water molecule to four others. Vast networks of bonds are forged this way at room temperature. The molecules link up to make a liquid, rather than move independently as in a gas. (The selfsame hydrogen bonds also make water blue, because they absorb a little red from sunlight, which contains all the colors of the rainbow.)

When water freezes, the hydrogen bonds hold each water molecule apart, at bond's length, so the structure of the solid is more open and less dense than the liquid. This is why we should blame hydrogen bonds for the loss of the *Titanic.*

Hydrogen bonding also plays a part in a puzzle that baffled scientists for years. Why is it that ice has a slippery surface that allows us to ski, skate, and slide so easily? The answer came recently from experiments by a team led by Peter Toennies of the Max Planck Institute for Fluid Dynamics in Göttingen, Germany. The scientists peppered ice with inert, low-energy helium atoms, which bounce off a surface so readily that they are exquisitely sensitive to the movement and arrangement of its topmost layer of molecules. The way that the helium atoms ricocheted from a single ice crystal cooled down to −243°C (−405.4°F) revealed that the tiny amount of available heat energy was able to stretch and compress hydrogen bonds between the water molecules at the surface by about 10 percent. One of the team, physicist Alexei Glebov, said that these findings suggest that at higher temperatures, such as those found in ice rinks, the topmost water molecules are bestowed with enough mobility to break and remake hydrogen bonds so that they can diffuse across the surface like marbles on a tray. That is also why ice remains slippery, even far below freezing.

Hydrogen bonds explain other mysteries of ice. Under atmospheric pressure and at temperatures below freezing, the water molecules link together in larger networks whose fundamental building blocks are "puckered" six-membered rings of hexagonal symmetry. This same symmetry is preserved in snowflakes.

Amazingly, there have been efforts to link together water molecules, one by one, to study the structure of water and ice crystal formation in formidable detail. David Clary, a theoretical chemist at University College London, and Jon Gregory of Cambridge University simulated the process using computer programs to model small, interacting groups of water molecules and "Quantum Monte Carlo" methods — in effect, a random way to test the shapes the clusters of water molecules adopt. Complementing their work, Richard Saykally of the University of California at Berkeley has created clusters of two, three, or more water molecules in beams that are cooled to near the lowest possible temperature.

Interestingly, the results of these experiments agree with Clary's predictions. "We have worked on clusters of two, three, four, and five water molecules, all of which join up to form flat rings," Clary says. "But when you add just one more molecule, then things change dramatically. With six water molecules you form a three-dimensional cage." A star is born: the smallest man-made snowflake.

Efforts to stick eight water molecules together have been made at Purdue University by chemist Timothy Zwier, post-doctoral associate Caleb Arrington, and graduate students Christopher Gruenloh and Joel Carney, who found that tiny ice cubes form spontaneously. In addition, they discovered that these cubes actually come in two forms, which have the same mass and structure but differ in the arrangement of the hydrogen bonds that hold them together.

Although the hydrogen bonds in the top layer of each cube were oriented in the same manner, the bonds in the bottom layer had one of two possible arrangements, with the bonds facing either the same direction or the opposite direction as the bonds in the top layer of the cube. "These findings verify what theorists have predicted for years; namely, that the eight water molecules preferentially form a cubic structure," Zwier says. "They also provide the first evidence that even in very small water clusters, water has the capacity to arrange its hydrogen bonds in several distinct orientations." This underlines the fact that the astonishing diversity of snowflakes emerges even when they consist of only eight water molecules.

You may be wondering how, in the name of Santa, Zwier and his coworkers could study these tiny cubes. They used an infrared laser to excite the clusters, causing the hydrogen bonds in the tiny cubes to stretch and contract.

By analyzing the wavelengths of the resulting spectrum and comparing their results with computer calculations from Kenneth Jordan's group at the University of Pittsburgh, they were able to determine the molecular arrangement of the hydrogen bonds within the cubes. "Since the two structures are virtually identical in energy, the configuration that a particular cluster takes depends on the specific collisions the cluster undergoes while it is being made," Zwier says. "It is interesting that already with only eight water molecules, water makes up two different 'phases' which differ only in the orientations of the hydrogen bonds." When it comes to the number of forms, or phases, of ice, water has more solid phases — nine total — than any other known pure substance because it can form phases which differ only in the orientations of the hydrogen bonds. No wonder every crystal is different.

Snow Birth

For snow to fall, you first need clouds. They form when warm, moist air is cooled, either by being pushed upward by flowing over a mountain, or when a wedge of cold air noses under a body of warm air in a frontal system. If the air is very clean, it can reach high levels of saturation with moisture — indeed, become supersaturated — before droplets of water appear. For droplets to form at lower levels, they require a nucleus to form upon. These are called cloud condensation nuclei.

As the air temperature declines, drops condense out of the water vapor with a diameter of about one-hundredth of a millimeter. Viewed from the perspective of any humans walking below, these vast congregations of tiny droplets form the pillows, tails, and streaks we call clouds.

Although the temperatures within a cloud are often well below freezing, the suspended water droplets will usually stay liquid. These droplets are supercooled and will remain in the air as a cloud unless it cools to extreme temperatures of −40°C and below, when they freeze into tiny crystals called diamond dust.

One manifestation of ice crystals high in the sky is wispy cirrus clouds. Another is optical effects that result from the crystals behaving like prisms, causing strange halos around the sun or bright spots called sun dogs.

For snow to form at temperatures higher than −40°C, a special particle called an ice nucleus is required. This acts as a base for the crystal. Such "snow seeds" are many and varied. Because there are so few ice-forming nuclei compared to the number of dust particles in the atmosphere, however, they are rather difficult to identify. They include soil, dust, bits of volcanic ash, or even material of extraterrestrial origin. (Each year some forty thousand tons of fine particles descend from the heavens. They are called Brownlee particles, after the scientist who first collected them in a U-2 spy plane, Donald Brownlee.) The snow seeds can even be added by human intervention: silver iodide crystals can be finely dispersed to induce snow to fall.

Compared to the number of droplets wafting around in a cloud, there are very few natural ice nuclei, although the actual number depends on temperature. For example, at −20°C, there is only about one nucleus per liter. You might think that, from this figure, we would know about the propensity of any cloud to form snow from the temperature alone. But nature, of course, is never quite this straightforward, explains Tom Lachlan-Cope, the British Antarctic Survey's "snow man." Many clouds at around −10° to −20°C contain many more ice crystals than would be expected from simple theory. These complex "mixed phase clouds" show that weather scientists still have some way to go to understand the physics of a seasonal sprinkling.

Snow Diversity

The snowflakes that float down from the sky are all different because of the way the tiny proto-flakes and their descendants attract new water molecules to their corners. As the crystals fall toward Earth through air of different temperatures and humidities, buffeted by winds, they grow in their own unique ways as water molecules attach to them.

No two snowflakes are alike, because their shape arises from the interplay between the random arrival of water molecules and their preference for assembling in a hexagonal fashion, as we saw earlier. The crystals devour water vapor in the cloud, growing rapidly until they are large enough to fall as snowflakes, the so-called Bergeron Findeson process.

Large-scale flake growth was modeled in a computer in the mid-1980s by Johann Nittmann of the Dowell Schlumberger company in France and Eugene Stanley of Boston University. They were building on work that dates back to at least 1949, when the Russian physicist G. P. Ivantsov came up with an equation to predict the shape of the tips of a flake. His efforts were modified to predict the correct rate of growth by teams at the University of Michigan/Schlumberger Doll and at the University of California, Santa Barbara.

"Each move to assemble a water molecule can be thought of as the spin of the wheel in Monte Carlo gambling," Stanley says. "Then the sequence of, say, the trillion molecules needed to make a snowflake corresponds to a trillion spins of the wheel."

A growing snowflake tends to take on more molecules at its tips, as the growth screens off less accessible surfaces toward the middle of the flake. "It is very difficult for a random walker [that is, a randomly moving water molecule] to walk along a narrow fjord in a snowflake without sticking on the side wall," Nittmann explains.

Using computer simulation, the team re-created random motions of up to twenty thousand molecules. The results surprised the researchers. "When we put the model on our computer, we got a picture which so strikingly resembled a real snowflake we were flabbergasted. No one had ever used a computer to generate a picture that really looked like a snowflake," Stanley says.

Atmospheric snow crystals come in the following forms: plates, stellars, columns, needles, spatial dendrites (in other words, branched), capped columns, and irregular crystals. For example, at around –1°C the ice crystals grow into thin plates. As the temperature drops to –11°C, they grow into hollow columns. At –15°C everyone's idea of a conventional snowflake emerges, with star patterns starting to grow.

Stanley and Nittmann found that by tinkering with the parameters of their computer model, they could make the two extreme forms of snowflake: the flat, hexagonal plates and the Christmassy variety with six feathery arms. The end result seemed to depend on winds. With a great deal of random gusts, it becomes possible to "fill in" the crystal so that hexagonal plates result.

Various potential careers await each ice crystal, depending on ambient air temperature and humidity. As they fall through the cloud, the crystals clump together by a process called aggregation, which involves the slight thawing of crystals. These wet crystals then collide and freeze again, forming snowflakes. Aggregation occurs best at temperatures around freezing. It does not occur if the temperature is too cold, because the crystals are dry.

The biggest snowflakes form when the temperature is between 0° and 2°C, but if the temperature is higher, the snow melts to form rain or sleet. This can make forecasting tricky, since a melting snowflake at high altitudes can appear to be a fat water droplet to a ground radar network. Clues to distinguish the two have been studied by *Snoopy,* the geriatric research aircraft belonging to Britain's Meteorological Office.

At times the ice crystals swirling through clouds — their collisions, breakup, and interactions with water droplets — can be literally illuminating. This activity generates electrical charges that result in radio static and even in lightning discharges. St. Elmo's fire, a brushlike discharge of atmospheric electricity that makes no sound, is often connected with the actions of ice crystals, snowflakes, and soft hail or graupel.

Snow Obsessives

The snowflakes that adorn Christmas cards are somewhat idealized. Most snowflakes are ugly sisters: only 1 percent are symmetrical. Nonetheless, the beauty and symmetry of snow have been apparent to people for millennia. The Chinese commented on this characteristic in 135 B.C. Europeans noted their symmetry at least by the Middle Ages, when the Dominican philosopher, scientist, and theologian Albertus Magnus wrote about snow crystals in 1260.

At the beginning of the seventeenth century, the same subject beguiled the German mathematician and astronomer Johannes Kepler. "There must be some definite cause," he wrote in 1611, "why, whenever snow begins to fall, its initial formation invariably displays the shape of a six-cornered starlet. For if it happens by chance, why do they not fall just as well with five corners or with seven?" In his pamphlet *On the Six-Cornered Snowflake,* Kepler draws parallels with honeycombs and the pattern of seeds inside pomegranates. But he was unable to explain the flakes' hexagonal form. This was also studied by the English scientist and inventor Robert Hooke, who published drawings in his scientific masterpiece *Micrographia* (1665), an account of his microscopic investigations.

The appearance of flakes was not widely appreciated, however, until the middle of the nineteenth century, when the book *Cloud Crystals,* with sketches by "A Lady," was published in the United States. The lady in question had captured snowflakes on a black surface and then observed them with a magnifying glass.

The end of that century saw perhaps the most famous snowflake obsessive of all. In 1880, at the age of fifteen, Wilson Alwyn ("Snowflake") Bentley, of Jericho, Vermont, was presented with a microscope by his mother. By his seventeenth birthday, Bentley had accumulated some three hundred drawings of snowflakes. His father stoked his son's obsession by purchasing a camera, exactly what Bentley needed to preserve his fleeting snow crystals forever. January 15, 1885, is the day credited as being the first date he made a successful photomicrograph of a snowflake.

After sixteen years of collecting snowflakes at an average rate of seventy to seventy-five per storm, Bentley had gained considerable knowledge and understanding of which storms, or parts of storms, and temperatures yielded the best subjects. The year 1931 saw the culmination of Bentley's lifelong passion, when his *Snow Crystals* was published. Around half of his 5,381 photographs were selected to illustrate the book.

On December 23, 1931, Bentley died of pneumonia. He contracted the illness when he walked six miles home in a particularly severe snowstorm.

The legacy of all these photographs and drawings is the idea that no two snowflakes are identical. Yet a few years ago, Nancy Knight of the U.S. National Center for Atmospheric Research found what appeared to be twin snowflakes, matching hollow hexagonal prisms. They may have *looked* very similar, but a precise match is so unlikely as to be impossible. The reason can be illustrated by thinking of the numbers involved. Ice has a density of about one gram per cubic centimeter, and from this figure, and the mass of a known number of water molecules, we can calculate that a one-centimeter cube of ice contains around 30,000,000,000,000,000,000,000 molecules, give or take a handful.

A snowflake, then, might contain on the order of 100,000,000,000,000,000,000 water molecules. Given Tim Zwier's discovery that the eight molecules in his nanocubes can be arranged two different ways, imagine the variety of ways that you could build a snowflake out of one hundred million million million water molecules. The short answer is, you can't.

Aside from Knight's near identical twins, there are many other exceptional flakes, notably in terms of their size. Most snowflakes are less than one-half inch across. Given near-freezing temperatures, light winds, and unstable, convective atmospheric conditions, much larger and more irregular flakes can form. At those times snowfalls deep and thick enough to impress the Bohemian king Wenceslas can occur.

No routine measurement of snowflake dimensions is taken, so there is no precise record of which flakes have been the biggest. There are, however, plenty of candidates. On January 10, 1915, huge snowflakes fell among the usual ones in Berlin. They measured up to four inches across. "Gigantic snowflakes," the *Monthly Weather Review* of February 1915 trumpeted. "They resembled a round or oval dish with its edges bent upward." Monster flakes some three and a half inches in diameter fell in Chepstow, England, in January 1887, according to a report in *Nature* by a Mr. E. J. Lowe. But the real mother of all known flakes fell on Fort Keogh, Montana, in the same month as the Chepstow flake. They landed near a ranch belonging to Matt Coleman and were described as "larger than milk pans" in *Monthly Weather Review.* It is claimed that they measured fifteen inches across and almost eight inches thick.

On the Ground

Even when snow has settled on the ground, there is no end to its dazzling diversity. Snow is a common Indo-Germanic word (German *Schnee,* Latin *nix,* French *neige*) that is used for both falling and lying snow. Living in the frozen North, however, the Eskimo, or Inuit, have so intimate a relationship with snow that they have coined many different words to describe its many varieties. Or so the story goes. This has been called the "Great Eskimo Vocabulary Hoax," for in fact the Inuit probably use only about a dozen words, similar to what is found in English (snow, sleet, slush, blizzard, avalanche, hail, hardpack, powder, flurry, and dusting). The linguist Geoffrey Pullum speculated that the supposed number of Inuit words snowballed, so to speak, this century because it "comports so well with the many other facets of their polysynthetic perversity: rubbing noses; lending their wives to strangers; eating raw seal blubber; throwing Grandma out to be eaten by polar bears."

Whatever you agree to call it, snow varies greatly in its water content. Ten inches of fresh snow can contain as little as one-tenth of an inch of water or as much as four inches, depending on crystal structure, wind speed, temperature, and other factors. The majority of U.S. snows, for example, have a water-to-snow ratio of between 0.04 and 0.10.

We can also tell something about thc nature of snow by the way it sounds. You can almost *hear* when the ground is cloaked by a heavy snowfall. A thick layer of fresh, fluffy snow readily absorbs sound waves. If the snow's surface becomes smooth and hard as it ages, or if it has been polished by winds, it will actually help reflect sound waves. Sounds may then seem clearer and travel farther.

The way snow looks is something of a puzzle. Snow is made up of ice, yet there is one property that sets it apart from its component crystals: its color. To explain why one is blue and transparent and the other brilliant white, we need to understand how they behave in sunlight.

Most natural materials absorb some light, which gives them their color — blue in the case of water, for instance. Ice also is blue because it absorbs light in the red and yellow part of the sun's electromagnetic spectrum slightly more efficiently than blue. Because the difference in absorption is so small, you need large quantities of ice, as in a glacier, to see its blue color. Clearly glaciers are made even bluer by all the air bubbles, dust, dirt, and small particles they contain. Light traveling through the glacier scatters and reflects off these imperfections. As it bounces around, the ice has more time to absorb red/yellow light, deepening the blue.

In contrast, snow appears white. The complex structure of

snow crystals results in countless tiny surfaces from which visible light is reflected. The light rays bounce from one crystal of ice to another so that the light propagates randomly but efficiently through a snowflake before finding its way out again. No absorption takes place because the light waves travel only a short distance through the material due to the efficient scattering process.

The ability of snow to reflect light is also the reason it is primarily melted by warm air rather than sunshine. The same light-scattering process goes on in white paint, fog, clouds, and the fluffy white beards of rubicund gentlemen riding on reindeer-drawn sleighs. When this scattering reaches a particular pitch of intensity, the light waves can interfere with each other in such a way that the light actually becomes trapped, a process called Anderson localization, which has been demonstrated in a powdered semiconductor by Diederik Wiersma of the European Laboratory for Non-Linear Spectroscopy in Florence. "You can think of it as 'semi-frozen light' which keeps on running around in random loops," he says.

Cosmic Snowballs

The appearance of snow can be deceptive, however. One extraordinary example was recently snapped by a camera aboard NASA's Polar Satellite: a cosmic snowball that broke apart 15,000 kilometers above the Atlantic. Perhaps the biggest snowball ever seen, it was not made by diligent schoolchildren or deranged atmospheric scientists, but seemed to be an alien affair, one of the house-size comets that pelt our planet at the rate of one every few seconds.

There is, however, bitter scientific controversy over whether such objects really exist. Many scientists regard as preposterous the suggestion by Louis Frank of the University of Iowa that so many impacts — thirty snowballs a minute, each weighing 20 tons, a cosmic snowballing equivalent to a million tons of water a day — could have been missed for so long. "Earth's sky would sparkle like a Christmas tree" if it were hit by a steady shower of snowballs, Bashar Rizk of the Lunar and Planetary Laboratory in Tucson, Arizona, says. He claims that each resulting twinkle would be as bright as the moon and last for a minute. At the time of this writing, the great snowball debate is still raging.

Snow History

Snowfall of a conventional kind can be put to intriguing uses. As snowflakes flutter through the atmosphere, they scrub it of dirt. One study at the University of Wisconsin at Madison found that dendritic stars are better than raindrops at gathering up pollutants. This might suggest that we should engineer snowstorms to scrub the skies of pollution. Putting this somewhat speculative idea aside, the scavenging properties of snow are also handy for those interested in recording the chemistry of the atmosphere, providing clues to Christmas past and Christmas future.

Fallen snow contains frozen relics of what was once in the sky. Slice through a layer, and you see a series of snapshots of the atmospheric chemistry and pollutants — the particles of desert sand, volcanic ash, radioisotopes from nuclear tests, and automobile pollution that the flakes have scoured from the skies.

The same is true for ice sheets or glaciers, which are formed from the compression of snow over many years and contain the chemical fingerprint of gases, acids, pollen, and dust. Evidence of ancient disasters also is trapped and frozen inside: traces of ammonium formate suggest forest fires, acid reveals volcanic eruptions, and radioactivity signals the fall-out of nuclear tests and accidents.

One team of scientists from the United States and China has reconstructed a detailed climate record for the past 130,000 years from a thousand-foot-long ice core they drilled into the Guliya Ice Cap, a 77-square-mile glacier sitting 22,014 feet high in the mountains of western China. "A record of this length from the subtropics is truly unprecedented," says Ellen Mosley-Thompson, professor of geography at Ohio State University.

The Ohio team cut their core sample into 34,800 pieces that were then tested for oxygen isotope ratios, dust, pollen, and nitrate, chloride, and sulfate ions. For years researchers have assumed that the climate in the tropics and subtropics has been fairly stable, Mosley-Thompson says. But the new core from Guliya, along with their other low-latitude ice core records, suggests that these regions may have experienced considerable climate variability during the past 100,000 years.

In the remote interior of Greenland, another vast stretch of atmospheric history is trapped in the ice sheets. The first half of this decade, the Greenland Ice Core Project, a thirty-member team from eight European nations, drilled a hole through the highest point on the Greenland ice sheet. This allowed the team to extract a core of ice that, if reassembled, would be close to two miles long, forming a record that stretches back more than 200,000 Christmases.

Among the scientists was John Moore, a glaciologist from Britain. "We drill about 2.5 meters each time the drill goes down," Moore explains. "It takes an hour to lower it down the hole, ten minutes to drill, and another fifty minutes to raise it." After initial analysis by scientists in the field, the remainder of the four-inch-diameter ice core was sent to Copenhagen for storage and more detailed analysis. The period covered by the compressed ice samples encompasses one ice age, an earlier warm period similar to the present, and part of the previous freeze.

Physical and chemical analysis of about eighty dissolved constituents reveals the history of the climate. "From the uppermost ice layers, we can see dust particles from the Chernobyl nuclear accident in 1986," says Joergen Taageholt of the Danish Polar Center. "Much further down, we have found traces of acid rain in the ice caused by volcanic activity when Vesuvius erupted in A.D. 79."

Crushing this ice releases the breath of ancient atmospheres. Trapped bubbles give evidence of concentrations of carbon dioxide and methane that may already have influenced patterns of climate change during the preceding millennia through the greenhouse effect. "We are trying to look back in the past to understand how climate is controlled and give us more confidence in predicting future climate," says David Peel, a program leader from the British Antarctic Survey.

Ancient temperatures can be deduced from the composition of oxygen isotopes. When ocean water evaporates in the low and middle latitudes, the water molecules contain a given proportion of the two "flavors" of oxygen, the heavy oxygen 18 and the light isotope, oxygen 16. But this proportion changes as clouds drift toward the poles. The heavier water molecules, which contain oxygen 18, condense faster than the light ones and precipitate sooner as rain or snow. By the time the remainder falls on Greenland, the oxygen 18 is depleted. How much, of course, depends on the temperatures en route.

These ice records also contain hints that climatic change may be quicker than we think, with profound implications for the future of white Christmases. It seems common sense to expect that the gradual buildup of atmospheric pollutants such as greenhouse gases will lead to an equally gradual change in global climate patterns. But this comforting idea has been exploded by climate records locked up in the ice, which reveal that during ice ages, the global climate can lurch dramatically from warm to cold and back again.

Climatologists had reassured themselves that these "climate flips" were linked to the ice ages and changes in ice sheets. In one scenario, armadas of icebergs were let loose, melting fresh water into the North Atlantic and thus disrupting the delicately balanced global ocean circulation system that today keeps Europe warmer than North America. Now, however, there is evidence that climate flips could occur in the absence of the ice sheets within a century or two — a mere eye blink in the vast expanse of geological time.

Traces of climate flips locked up in Greenland ice cores suggest that Earth's climate cools significantly and abruptly in a naturally occurring 1,000- to 3,000-year cycle. The studies of isotopes of oxygen — an indirect measure of temperature — have revealed these climate flickers during the final part of the last ice age, between 10,000 and 30,000 years ago. The temperature record shows square-wave-shaped fluctuations, with a period of relative stability followed by a rapid rise or dip in temperature. However, the same record has been relatively flat and boring over the past 10,000 years, which might mean that we can sleep soundly in the knowledge that climate flips occur only when Earth is in the grip of ice. Or can we?

Evidence that these alarming flickers also took place recently, long after most ice sheets had disappeared and conditions closely resembled those of today, has come from work by Gerard Bond, a paleoclimatologist at Columbia's Lamont-Doherty Earth Observatory in Palisades, New York. Studying layers of rock fragments transported by glacial icebergs and sea ice to the North Atlantic, subsequently deposited on the sea floor, and then buried by sediments near Iceland and Greenland, he found a number of abrupt coolings that took place during the Holocene — the term geologists use to describe the past 10,500 years, after the last ice age ended and human civilization began to flourish.

The regularly spaced layers of debris showed that the amount of floating ice — and thus the global temperature — peaked every 1,000 to 3,000 years. Bond dated the greatest amounts of this debris at 12,300; 10,800; 8,000; 5,700; 3,900; 2,750; and 800 years ago. This corroborates earlier work by Suzanne O'Brien and colleagues from the University of New Hampshire, who likewise found evidence, in a study of ice cores drilled in Greenland, for a series of flips in the atmospheric circulation during the Holocene in the high northern latitudes. One of these events, a sharp cooling around 8,200 years ago that lasted for about 400 years, seems to have been connected with a pattern of unusual climatic events that affected many parts of the Northern Hemisphere. "The abrupt coolings in the Holocene era are not as great as those that occurred during the ice ages but still might be significant enough to cause severe winters, agricultural disruptions, and other adverse impacts on people," Bond says.

The most recent cooling cycle might have been the Little Ice Age, which began around 1100 and peaked a few hundred years later. Glaciers covered the Alps, Alaska, New Zealand, and Sweden; snow blanketed Ethiopia's high mountains; global temperature was 1.1°C (2°F) lower than now; and Europe and North America suffered severe winters. "If this is indeed a regular climate rhythm, it is still going on today," Bond says. "By understanding what causes these sudden climate changes when Earth is relatively free of ice, we can anticipate the next event. The odds of a future climate jolt could be higher than we thought."

But that leaves one huge puzzle: what causes these climate flips? Perhaps they are triggered by clocklike changes in the way the oceans circulate. The oceans are a crucial engine of climate, transporting vast amounts of heat energy around the globe. Another possibility is that the climate is responding to some external factor, perhaps the amount of the sun's radiation beating down on Earth. Recent findings have linked a 100,000-year climatic cycle to the influx of cosmic dust, the particles that help seed snow in the first place.

The Christ Child

Christmas itself is connected with one well-known climate cycle whose effects are felt every three to seven years. A huge region of warm surface water shifts toward the west coast of South America, affecting global rainfall and winds. The warming can be detected by a grid of buoys and sensors, called the Tropical Atmosphere Ocean Array, which measures atmospheric wind and ocean temperature. The popular name of this warming is El Niño, "the Christ child," because it usually coincides with Christmas.

This benign name belies El Niño's impact, which is of biblical proportions. Formally called the El Niño/Southern Oscillation, it directly affects the climate of more than half the planet. Trade winds in the tropical Pacific Ocean slacken and reverse directions. The pool of warm surface waters that shifts from one end of the equatorial Pacific to the other is accompanied by a great center of tropical rainfall. On average, 10 percent more rain falls than usual.

The onset and intensity of Indian monsoons and African downpours are influenced by El Niño, as are the frequency, severity, and paths of storms in the Pacific; the occurrence of regional droughts, forest fires, floods, mudslides, and hurricanes; and the movement of shoals of anchovy, tuna, shrimp, and other commercial fish. The latter is good news for some fishermen, bad for others: the latest manifestation (1997–8) saw the loss of millions of adult sockeye salmon in the Bering Sea and large catches near San Francisco of the equatorial game fish mahimahi.

El Niño also affects the jet stream across America, causing widespread flooding. Dramatic examples include the huge Mississippi River floods of 1993 and the destructive California rains in 1994–5. During the 1997–8 El Niño the temperature rises in the Pacific Ocean off the Great Barrier Reef and Panama caused coral bleaching, when algae essential to the survival of the reefs was expelled. Nor is the Far East immune to the effects of El Niño, as underlined by the smogs that cloaked huge areas of Southeast Asia when fires burned out of control early in 1998.

In 1982, an intense El Niño was not predicted and not

even recognized during its early stages. Eventually it caused thousands of deaths and more than $13 billion in damage worldwide. El Niño is second only to the seasons themselves in driving worldwide weather patterns.

But El Niño also brings with it a gift — one that is causing tremendous excitement: the prospect of long-term climate forecasting. "Our models cannot predict that it is going to rain the day after Christmas next year, but they can predict whether it will rain more than usual next December," says Nicholas Graham, a climatologist at the Scripps Institution of Oceanography in La Jolla, California.

"We will not be able to say there will be a drought, but rather that there is a good probability of a drought occurring in a specific region of the world," adds Mark Cane of Columbia University's Lamont-Doherty Earth Observatory.

In 1986 Cane and colleague Stephen Zebiak put together the first computer model that successfully predicted El Niño. At about the same time, Graham and fellow Scripps scientist Tim Barnett developed a novel method that linked El Niño to the tropical Pacific to the global climate system, an advance that allowed them to use Pacific sea surface temperature information to predict rainfall, temperatures, and other climatic conditions elsewhere.

Almost a decade later, El Niño was used to predict corn harvests six months in advance and halfway around the globe. The link between El Niño and corn yield in Zimbabwe, one of the ten nations in southern Africa where it is the major food crop, was uncovered by Cane and Gidon Eshel of Lamont-Doherty and by Roger Buckland of the twelve-nation Southern African Development Community's Food Security Technical and Administrative Unit in Harare.

The scientists demonstrated that between 1973 and 1990,

low rainfall and poor harvests fluctuated nearly in step with the periodic warming of eastern equatorial Pacific surface waters. Ample rainfall and corn harvests corresponded with the opposite extreme of the El Niño cycle, in which eastern Pacific waters become unusually cool.

Computer models that predict El Niño in August of one year can provide indications of probable crop yields the next April. This advance warning allows farmers to change strategies and gives policymakers time to conserve water, import food, or take other precautions.

The work of this team suggests that many links between El Niño and crop yields in other parts of the globe are still to be uncovered. This Christmas present holds so much promise that Columbia and Scripps created a research institute to provide early warning of El Niño and other climate variabilities that influence droughts, floods, and a range of destructive weather patterns worldwide.

Snow Dreams

Given everything that scientists now know about the vagaries of the climate, the molecular structure of ice, the fractal structure of snow, and the origins of flakes in clouds, can they engineer that white Christmas that we all dream of? Thanks to modern technology, it is now possible to summon snow without the help of the weather. At ski resorts during a mild winter, water is shot out through a cannon as a fine spray to produce snow. But this alone is not enough. As explained earlier, ice crystals need something to form around — they don't just crystallize in thin air. Rather than relying on dust floating in the atmosphere, the snow-makers use harmless bacteria for the job. The resulting snow is not as good as the real thing, and tends to be lumpy.

There are other methods of making artificial snow. Thanks to a facility built at the Suga Test Instruments Company in downtown Shinjuku, Japan, snow can even fall on the hottest July day. Unlike the more primitive snow made for ski slopes, the $1.25 million computer-controlled Suga simulator facilitates the creation of natural snowflakes. The company specializes in the study of the weathering process, and sees artificial snow as a valuable supplement to the combinations of sunshine, wind pressure, and rainfall used to put materials through their paces. Lab-made snow can also give insights into the atmospheric conditions that lead to the real thing.

Beyond this artificially produced snow, there is, of course, the *really* artificial variety for those who want to create an off-season winter wonderland. The traditional plastic snow — the kind made in Hollywood — consists of tiny flakes of white polyethylene that have to be collected and disposed of afterward. Now there are alternatives. Sturm's Special, of Lake Geneva, Wisconsin, has developed biodegradable flakes of modified polylactic acid polymer from corn or cheese by-products. The flakes are of random size, mostly irregular, and platelike.

Dietmar Voelkle and Frithjof Baumann of the Fraunhofer Institute for Chemical Technology in Pfinztal, Germany, have developed two new kinds of artificial snow. The first is biodegradable artificial snow, being made of foamed potato and cornstarch. It is ideal for Christmas window displays, say its inventors, because it sticks to anything that is damp. "One can even model icicles or build a snowman with the stuff," Baumann explains. "Add more water, and the flakes dissolve, so they also can be used outdoors without the need for sweeping up."

The second of their artificial snow creations is made of fire-resistant foamed polyethylene and is both more realistic and cheaper than the standard plastic snow. Depending on how a mill is set up for its production, it can be made to flutter and sparkle in a realistic way for use in the theater. This "glitter snow" has already made its stage debut in a performance of *Prince Igor,* at the Badisches Staatstheater Karlsruhe, that required "Siberian" snow.

Dreaming of an Artificial Christmas

There is as yet only one way to guarantee a white Christmas: head for the snow line, the boundary at altitudes beyond which you find perpetual snow. In high polar latitudes, for example, the snow line is at sea level. In northern Scandinavia it is somewhere around 1,200 meters, in the Alps at about 2,400 meters, and in the Himalayas at about 4,500 meters. But given all that we know about snow, surely we no longer have to be slaves to geography and season. Surely it should even be possible to provoke the weather to provide a white Christmas on demand.

With the kind of resources that would make Croesus wince, meteorologists say that it should be possible to engineer the atmosphere to guarantee a good sprinkling of snow. The trick is to reproduce the effect that triggers snow in the first place, when stationary cold air wedges under a moving mass of warm, moist air and lifts it so that snow condenses out.

In Britain, for example, an influx of warm, moist air flows

in from the west after a cold spell lasting for a few days, and violà — snow. For fun I asked the Meteorological Office in Bracknell what could be done to make absolutely sure that such conditions took place on Christmas Day. In London, for example, a Christmas Day snowfall is a rare event, occurring only once every twelve years or so.

"You would have to make Ireland much more mountainous — about the size of the Rockies — and extend it southward to form a block in the Atlantic," said a somewhat surprised spokesman. The resulting breeze from Europe would produce sufficient cooling over a week or so. "Then you have to make your block in the west fall over to let something back in, allowing weather systems from the Atlantic to ride over the cold air," he said. Somewhat unnecessarily, he pointed out that to erect and then dismantle a mountain range in Ireland made the scheme somewhat impractical.

Similar hypothetical plans have been under way in America to ensure that it would snow from Georgia to New Jersey on Christmas Day. This jolly gedankenexperiment was carried out by Allen Riordan, an expert on forecasting severe weather, who is the coordinator of the Southeast Consortium for Severe Thunderstorms and Tornadoes, located at North Carolina State University.

First, Riordan said, he would need some serious manpower. The Army Corps of Engineers and the Canadian government would have to work together to steer the Santa Claus flow — a Siberian cold air mass — over New England just in time to meet a warm and moist low-pressure system coming off the Gulf Stream.

Getting the cold air mass to descend over New England instead of the Midwest, where it normally lingers, is the most difficult part of his "project." The Army Corps of Engineers and Canadian government teams in the Canadian Rockies would have to work around the clock to enhance the mountains' height, which should ensure a successful rerouting of the Santa Claus flow. "It's going to take at least a couple of dump trucks," Riordan admitted.

To encourage the cold air farther east, the Canadian government would have to freeze Hudson Bay around a week before Christmas. The ice would maintain the cold air by keeping a high-pressure system on the land's surface. Then, as the cold air headed southward to the east of the Appalachian Mountains and over Virginia, it would meet warm, moist air coming over the offshore waters of the Gulf Stream.

To ensure that the warm and cold air masses collided, you would have to spark the formation of a low-pressure center along their common boundary. To accomplish this, the U.S. Navy would need to set off a series of massive explosions to create fireballs large enough to heat the air column and get the low-pressure system started.

"Then it would go by itself," Riordan said. The warm air would head northward over the Carolinas, the cold air would push south under it, and snow should be underfoot on Christmas morning.

Without Riordan's intervention, the real chances of snow falling in central North Carolina are "very, very slim," according to the National Weather Service in Raleigh. By Riordan's calculations, the best place for the explosions would be the Florida coast, near Jacksonville. A small price to pay for a white Christmas?

9

CHRISTMAS SPIRIT

Hot Tom and Jerry is an old-time drink that is once used by one and all in this country to celebrate Christmas with, and in fact it is once so popular that many people think Christmas is invented only to furnish an excuse for hot Tom and Jerry. . . . Good Time Charley and I start making this Tom and Jerry early in the day, so as to be sure to have enough to last us over Christmas, and it is now along towards six o'clock, and our holiday spirit is practically one hundred percent.

DAMON RUNYON, "FURTHERMORE"

THE APPEARANCE OF THE FIRST CHRISTMAS CARD was marked by controversy. There was criticism that this seasonal message of goodwill actually encouraged degenerate behavior. Look closely at one of the hand-colored versions, and you can see a brazen case of underage drinking: a girl in a green dress taking a gulp of red wine.

The complaints of the temperance movement and Victorian do-gooders had little effect, however. When Christmas cards became popular a couple of decades after the first was designed by John Horsley, drunken celebrations were still regularly depicted. An issue of the magazine *The Studio* sarcastically asked in 1894, "If we investigated all the cases of drunkenness in all these years, could we find a single one remotely traceable to this design of Mr Horsley's?"

Even today, Christmas celebrations would not be the same without the attentions of those diligent microbes that, through the process of fermentation, have provided us with alcohol for several millennia. We all have a favorite microbial waste product.

Being something of an oenophile, I like to savor the bouquet of a good claret. However, some of my friends prefer to gulp down a cold beer in hot pursuit of the bottom of the glass. Whatever your choice of libation, you are likely to consume more than usual during the Yuletide season.

Scientists have now put together a detailed picture of the molecular events that take place within the body of a seasonal reveler, from the first sniff of a wine's delicate bouquet and the way alcohol is absorbed into the bloodstream, to the resulting disruption of brain chemistry and the last groan of a hangover.

Christmas and Alcohol

In Chapter 5, I used Whistlecraft's rhyme to illustrate the excesses of seasonal feasting, from slices of "fatted beeves" to gorging on apple pie and custard. However, I omitted there what Whistlecraft's poem suggests should be used to wash down all this food. He provides an eye-watering, headaching list of possibilities: "And therewithall they drank good Gascon wine,/With mead, and ale, and cider of our own,/For porter, punch and negus were not known."

Throughout the centuries mead, ale, and wine have been consumed in copious quantities during seasonal festivities.

In Britain, for example, Christmas specialties were church ale, a strong brew, and lamb's wool, a hot concoction of mulled beer with apples bobbing on its surface. The latter was a wassail drink, a shared beverage that has forged bonds between revelers since Anglo-Saxon times.

The word *wassail* comes from an Old English salutation ("Be well"); it denotes Christmas merrymaking that involves carol singing and drinking the health of one's neighbors. The custom took various forms. In one New Year's Eve ritual, young women carried a wassail bowl of spiced ale about the local parish. They went from door to door, singing the following verse, presenting the drink to the inhabitants of any house they called on, expecting a small gratuity in return:

> We have got a little purse
> Of stretching leather skin
> We want a little of your money
> To line it well within.

In another variation, an empty wassail bowl was carried around by carol singers and a drink extorted at each door. The wassail bowl was also used in the home, sometimes with toast floating in it. This may be why we "toast" someone, although other explanations for this term have been offered.

Ritual alcoholic abuse of the body has gone on for thousands of years, thanks to one of the most ancient techniques of biotechnology — the fermentation of fruit and grain by the activity of fungi called yeasts. The result is both literally and figuratively a dizzying array of fluids, all of which contain alcohol (from the Arabic *al koh'l*), also called ethyl alcohol or ethanol, the correct chemical name for this molecule.

The yeasts on grape skins, for example, give wine an alcoholic kick by fermenting the sugars in grape juice. This biochemical process offered two attractions to the people who discovered it in prehistoric times. First, like cheese, wine can be safely consumed even though it is a partly spoiled food, and it resists further deterioration. Second, the alcohol produced by this specialized spoiling can be tolerated only in small amounts, giving alcohol a preservative effect. As food science writer Harold McGee points out, it was "alcohol's intoxicating power rather than its antiseptic properties that made wine and beers so popular so early."

Studies of how yeast performs this feat have fascinated and befuddled scientists throughout the ages, spawning the science of microbiology and laying the foundation for modern biochemistry. The first microorganisms to be isolated in pure cultures were beer and wine yeasts, and the word *enzyme,* which denotes the huge proteins that cells use to transform other molecules, was coined from the Greek words for "in yeast." This etymology seems very appropriate. By 700 B.C., Homer's time, wine had already become a staple beverage in Greece.

A Brief History of Booze

Alcohol has been an integral part of seasonal celebrations — the planting of crops, harvests, the inundation of the Nile in Egypt, and so on — for millennia. If Christmas seems to be marked by excessive drinking, remember that the feasting and drinking of the "renewal festival" for a pharaoh could go on for months.

As early as Neolithic times, people were familiar with the misery of the morning after. Residues of liquid in a jar found in Hajji Firuz Tepe, a Neolithic village in the Zagros Mountains of Iran, show a wine similar to what we call retsina today. Radiocarbon samples in the vicinity of the jar date it between 5400 and 5000 B.C. This means that people have been drinking wine — and celebrating with it — for two thousand years longer than previously thought.

The yellowish residue at the bottom of the jar was analyzed by a team led by Patrick McGovern of the University of Pennsylvania Museum using spectroscopic methods, which measure the light absorbed at various infrared and ultraviolet frequencies. Tests revealed that the residue contained the calcium salt of tartaric acid, which occurs naturally in large quantities only in grapes. The residue also contained resin from the terebinth tree, which was used in antiquity as an additive to inhibit the growth of bacteria.

Because wine was used in early medical formulas, the find also provided insights into the antique pharmacopoeia available in Egypt and Mesopotamia, and later in Greece and Rome. A second Hajji Firuz jar with a reddish deposit on its interior also was found to contain a resinated wine, which may turn out to be the red to go with the white.

Less clear is what the wine would have tasted like, though we can make an educated guess. The terebinth's lesser-known name is the "turpentine tree." "That's not exactly the kind of taste or smell that you'd necessarily want," McGovern observes. "But I think it would dominate."

Beer also has been available as a seasonal libation for many thousands of years. Perhaps even before the grape was domesticated, ancient people had found that starchy grains — barley, wheat, millet, and corn — could be fermentable. The trick is to allow the grain to sprout, at which time enzymes chew up the starch molecules and release their simpler component chemical units, notably glucose. In nature this process provides the sprouting seedling with an energy supply, but in brewing it supplies the glucose for yeasts to feed on.

McGovern's analysis of the fourth-millennium B.C. jar from the site of Godin Tepe, in the Zagros Mountains, suggests that beer brewing may have been carried on in Mesopotamia. We know for certain that barley and wheat beers were being brewed in Egypt as long ago as the third millennium B.C.

Both bread and beer were consumed by all members of Egyptian society, even supposedly the gods, and at virtually every meal. It is a wonder the ancient Egyptians ever managed to build in straight lines. The conventional view of ancient Egyptian brewing, based on records dating to around 1800 B.C., suggested that malted grain was preserved by baking it into a flat bread, which was then soaked in water and fermented when beer was to be made.

A few years ago, however, brewers from Edinburgh joined archaeologists in backing an expedition to Egypt to uncover the beer-making secrets of the pharaohs. The efforts to recreate Tutankhamen's tipple were carried out in a partnership between the Egypt Exploration Society and Scottish and Newcastle Breweries, led by Barry Kemp of Cambridge University.

The destination of the four-year project was Tel el Amarna, an ancient capital of Egypt halfway between Cairo and Luxor and once home to Tutankhamen (who died at the age of eighteen in about 1340 B.C.); his father, King Akhenaton; and Queen Nefertiti. Expedition member Delwen Samuel set out to excavate Tutankhamen's brewery and to find new evidence for ancient brewing using archaeological evidence rather than relying on traditional sources such as hieroglyphics, tomb models, paintings, and classical writings. She believes that previous work has misinterpreted the hints of ancient recipes suggested by paintings. Instead, she took a different approach, deducing the ingredients and brewing method by examining actual samples of bread and beer found in and around the tombs and especially from domestic areas where everyday beer was made. The beer itself is preserved in the form of solid residues left after evaporation. Most of these residues were recovered from rubbish dumps, where vessels had been discarded after breakage.

Charred grains of emmer wheat, a very rare species still used in a few Mediterranean countries, and barley were among the bread molds and broken brewing jars found in the remains of the temple bakery-brewery. Using microscopy, Samuel determined that funerary bread was made mostly from emmer wheat, occasionally flavored with figs, dates, or coriander, but that the grain used for beer was more varied, usually consisting of barley but also emmer and sometimes a mixture of the two.

Because of the remarkable preservation of some of the desiccated foods, Samuel was able to analyze the structure of starch granules in the remains of bread and beer, which provide important clues to baking and brewing methods. The conclusion: The ancient Egyptians employed a sophisticated two-step brewing process resembling some African methods still used today.

The grain was divided into two batches. One was made into malt by sprouting and gentle drying, then coarsely ground. The other batch may have been malted or not, but it was coarsely ground and well heated in water. The two batches were then mixed together. The active enzymes from the uncooked malt broke down the starch in the cooked grain, which was much more susceptible to enzyme attack. This was an excellent system for producing large quantities of sugar from grain in a system without thermometers or other means to control the processes closely, Samuel notes. The chaff was then removed by sieving, and the resulting sweet, cloudy liquid was inoculated with yeast and fermented. "The ancient Egyptians were much more sophisticated brewers than has been thought," Samuel concludes.

The National Institute for Agricultural Botany in Cambridge, England, grew enough emmer so that researchers could attempt the next step: re-creation of the Egyptian recipes. The bread made from this grain, using an ancient recipe, was very palatable, Samuel says, with a "distinctive nutlike flavor." The resulting beer, made by Scottish and Newcastle Breweries using modern equipment, turned out to be much stronger and spicier than modern beer. It was put on sale at Harrods for £50 a bottle under the label "Tutankhamen's Ale." The first batch has long since sold out, with the profits going to further research, but it may be possible to persuade the brewery to make ancient Egyptian beer a regular, if small-scale, brew.

Intriguingly, in societies where both wine and beer have been available simultaneously, beer has always been the drink of the commoner and wine that of the elite. This may explain why people are more inclined to celebrate a special occasion such as Christmas with wine.

Harold McGee points out that grain, the raw material for beer, is cheaper than grapes and can be stored for a long time in raw form before being brewed; its fermentation also is easier and less drawn out, so that "to the Greeks and Romans, beer remained an imitation wine made by barbarians who did not cultivate the grape." The same point was made around A.D. 230 by the Greek writer Athenaeus when he wrote on the invention of beer. Wine, he said, was first known and loved by the Egyptians, "and so a way was found among them to help those who could not afford wine, namely, to drink that wine made from barley."

From Lip to Blood to Brain — And Beyond

What goes on in our bodies when we drink during that Christmas party? After a couple of glasses of wine, alcohol produces a general feeling of well-being. Another glass, and inhibitions may begin to fall. Consume another couple of glasses, and you had better not drive. Once a bottle has been consumed, you feel unsteady on your feet. Two bottles, and you are drowsy and confused. Three produce a drunken stupor, and four can make an average man dead drunk — even dead, if consumed quickly enough. The following sequence of events will vary in detail depending on an individual's genes and body weight, but it is a reasonable guide to the passage of liquor through the body of anyone who enters the seasonal fray with gusto.

The First Sip

The bouquet, whether King Tut's ale or a fine Pomerol, is distinguished by the combination of taste and the more discriminating sense of smell, which results from the volatile vapors that pass into the nose. These arise during fermentation, when not only alcohol but also several hundred other substances are produced.

When you sniff a glass of wine, about ten trillion odorous

molecules waft over a yellow membrane at the top of the nose. This apparatus, which is plugged directly into the brain (where the smell center is close to the limbic center, the seat of the emotions), translates the bouquet into electrical activity. Although the details are not well understood, a great deal of progress has been made in unraveling how we can distinguish the smell of a savory Christmas turkey from that of the awful aftershave given by a distant relative.

Smell is the most enigmatic of our senses. At one time it was a mystery why we can smell anything at all. How could odor molecules, which are water-repellent, pass through the moist cell walls in the nasal membrane? Then a team at Johns Hopkins Medical School in Baltimore found a family of proteins that cart smells into the cells. Once inside, each chemical component of the wine vapor, for instance, is thought to lock onto molecular docking sites, the first stage in sending a signal to the brain that can identify a wine as coming from the sunny side of a Burgundy valley or California.

Scientists used to think that the smell of a chemical depended on the shape of its component molecules, which in turn affected how the molecules docked with the "smell" sites located within the nose. But Luca Turin of University College London believes the shake, rattle, and roll of the molecules are a more significant factor. This revives an idea that goes back sixty years, when a British chemist named Sir Malcolm Dyson proposed that the nose might work like a spectroscope, a machine that detects molecular vibrations by shining light of different wavelengths on a molecule and seeing which ones are absorbed.

Molecules with very similar shapes can have different vibrations and thus different smells, Turin explains. One molecule, shaped like a hamburger, smelled quite different depending on whether the "burger" component was an iron or a nickel atom, for example. By the same token, boron and sulfur compounds of different shapes can smell the same because sulfur-hydrogen bonds and boron-hydrogen bonds vibrate at the same frequency.

Perhaps the most elegant, and strongest, demonstration that we smell vibrations has come from observing the effect of exchanging hydrogen in the molecule acetophenone for its heavy brother, deuterium. That leaves the molecule with the same shape, but the new vibrational frequency of the carbon-deuterium bond is much lower than that of the carbon-hydrogen bond. As a result, acetophenone and its deuterated version smell quite different.

The fragrance of a fine wine or beer results from the interaction of hundreds of different molecules in the air with receptor cells in the nose. Their signals travel first to the brain's "olfactory bulb" for preliminary processing and then to the cerebral cortex, which interprets the pattern as a particular smell. How do we recognize more than ten thousand different odors? If each nerve carried a receptor tailored to a specific odorant (smell molecule), then the brain would know what it was smelling simply by knowing which nerves were active. But to give the full gamut of aromas, from cinnamon to roast turkey, there would need to be as many types of receptor as there were smells. On the other hand, there could be just a few types of receptor, each reacting in a slightly different way to a given smell. The brain would have to compare lots of messages from this handful of receptor types to work out what it was smelling. Our eyes work in a similar way: they have only three types of color receptors. By comparing the responses of red, green, and blue receptors, the brain paints relatively few nerve signals into all the hues of the spectrum.

Scientists now think that we have something between fifty and a thousand different smell receptors, which stick out from nerve cells in the nasal cavity to dock with airborne molecules. Recent work by Stuart Firestein and colleagues at Columbia University has even started to link the receptors with particular smells. The first aroma to be matched was octanal, which smells like meat to some and has a light citrus odor to others. A good fit between octanal and receptor changes the shape of the receptor, triggering a cascade of processes that send information to the brain. That still leaves the problem of how the response of these receptors to the complex mixture of chemicals in an odor is translated into a characteristic pattern of brain activity.

Enter the bee. "What we can smell, honeybees can smell. There is not a big difference," explains German scientist Randolf Menzel. This is good news for smell research, for there are plenty of things you can do to a bee that you dare not do to a human.

Menzel wanted to know what goes on in the insect's brain when it sniffs a carnation. The smell of the flower has about thirty components, so what appears in the brain — a single pattern of activity or thirty different patterns?

Menzel, Jasdan Joerges, and colleagues from the Free University of Berlin uncovered the code the brain uses to represent each smell by studying the map of activity in the bee's olfactory bulb. This first stage in the pathway by which the brains of both bees and humans process smells looks like a bunch of grapes, each "grape" consisting of a bundle of nerves called a glomerulus.

Their work showed that odors set up patterns of activity in the glomeruli; the more intense the smell, the greater the contrast in activity. Each pattern represents an odor, and each odor evokes a specific pattern of activity in the glomeruli. But, most exciting of all, the group established that the smell of a carnation is not simply the addition of the patterns made by each of its thirty component smells. Rather than being flooded by patterns of all thirty components, the bee experiences a single pattern of activity, and it is that pattern that corresponds to the smell of a carnation.

When, for example, two basic chemical smell components were mixed, the individual patterns of nerve activity combined but also included features unique to the combination, in parts of the glomeruli where the component patterns had inhibited each other. "The patterns are predominantly a simple addition but have subtle differences," Menzel says. "If you mix more than two, say three or four, then a unique pattern develops." Something similar is going on when we sniff a glass of wine. It should be no surprise that different vintages of the same wine can create different patterns of activity in the brain.

Taste is crude by comparison with smell. Called chemoreception, it takes place in nine thousand taste buds on the tongue. The molecules that evoke taste are called sapid molecules (from the Latin *sapere,* meaning "to taste"), and in general they have to be soluble in water if they are to penetrate the taste buds. Four types of taste buds respond to the saporous units in a drink: sweet (found on the tip of the tongue), salty (middle front, on either side), sour (middle back, on either side), and bitter (back).

Into the Bloodstream

After we take the first gulp of a drink, it passes down the gullet and into the stomach, where it can be absorbed into the bloodstream through the stomach walls, which are mucous membranes. The alcohol is taken up at different rates depending on the drink: a warming toddy, such as glogg or glühwein (mulled wine), will intoxicate more rapidly than a chilled wine, because the heat stimulates the blood supply to the mucous membranes, allowing alcohol to be absorbed more quickly into the bloodstream. Sparkling alcoholic drinks, such as a gin and tonic, also are more intoxicating, because carbon dioxide bubbles hasten the passage of the alcohol into the small intestine. Alcohol is absorbed about three times faster there than in the stomach. That is why champagne "goes to one's head."

Cell Biochemistry

A compound produced by alcohol-soaked brain cells can inhibit the release of messenger chemicals in the brain. In this way the chatter between the nerve cells within the brain of a Christmas reveler is damped down by alcohol, according to researchers at the Washington University School of Medicine in St. Louis. "I'm hopeful that we can now figure out a thousand-year mystery," says Richard Gross, one of the team. "Despite the structural simplicity of alcohol, nobody understands the biochemical mechanisms that are responsible for its neurological effects."

In a 1996 study published in the *Journal of Biological Chemistry*, Gross and Rose Gubitosi-Klug linked potentially intoxicating changes in brain chemistry to the production of fatty acid ethyl esters, a group of compounds that are made when alcohol combines with fatty acids. "Basically, we came up with the concept that it was a chemical in the process, not the ethanol itself, that was the active agent," Gross explains.

The team studied the effects of alcohol on genetically engineered insect cells and cells taken from a part of the brain called the hippocampus, a structure involved in memory. The compounds apparently speed up the release of charged potassium atoms from brain cells, which in turn inhibit the release of messenger chemicals called neurotransmitters. That interferes with the communication between brain cells by making it difficult for cells to absorb enough calcium to trigger the release of the neurotransmitters. A slowdown in neurotransmitter release could lead to slurred speech, clumsiness, slow reflexes, a loss of inhibitions, and short-term memory loss.

Into the Brain

While biochemists continue to investigate the effects of alcohol on individual cells, body scanners have revealed the havoc that an alcoholic drink can wreak across regions of a wassailer's brain. One such study was carried out by Malcolm Cooper, John Metz, and colleagues at the University of Chicago using a scanning method called positron-emission tomography (PET). They used a glucose molecule labeled with radioactive fluorine to highlight, under the gaze of an array of detectors in the scanner, the brain locations that had the greatest hunger for the energy molecule and thus the highest metabolic activity.

In one double-blind trial, the researchers found metabolic differences between the brains of those who had imbibed alcohol and those who had not. The former had more activity in the left side of the brain, and in particular the frontal and temporal lobes. The scanner also revealed a spectrum of effects that probably are not too surprising to anyone lurching out of a Christmas party: a surge of metabolic activity in the speech area (suggesting why drunks talk too much); a surge in the regions of the cerebellum, which coordinates movement (which is why alcohol makes us stagger); and a surge in the limbic system, a region that controls primitive responses such as sexual arousal and violence. This might be related to the typically boisterous behavior of heavy drinkers.

The Age Factor

The young are more likely to be able to consume more alcohol, and to become rowdy, than older people because it is less likely to make them feel sleepy. That is, of course, if young people behave anything like young rats. Junior rodents are more sensitive to alcohol-induced learning and memory deficits than adults. At the same time, alcohol does not make them feel as sleepy — potentially allowing them to drink more than adults and thus causing more harm — according to studies by Scott Swartzwelder and colleagues at Duke University Medical Center and the Durham Veterans Administration Medical Center in North Carolina, reported in the journal *Alcoholism: Clinical and Experimental Research.*

"We've shown that the developing brain has exactly the wrong combination of sensitivities to alcohol," Swartzwelder says. "It's capable of staying awake through a prolonged bout of drinking, but at the same time, it is sustaining far more damage to memory and learning systems than an adult brain receiving an equivalent amount of alcohol."

Scwartzwelder's studies suggest that as few as two drinks could inhibit learning and memory in a young person but would have little effect on an adult. Such damage is occurring at a period in development when the brain is most receptive to learning. At no other time can the brain absorb and retain so much information as during childhood, and alcohol potently depresses that ability.

The first preliminary study to follow up this find — on 21- to 30-year-olds — did indeed suggest that after only one drink, memory performance correlates with age. "The memory performance while on alcohol was better among the older people in the sample than in the younger ones," Swartzwelder says. Perhaps this is one reason childhood Christmas memories are more vivid than those from adolescence, when many people begin to experiment with alcohol.

Expulsion

The body treats the alcohol from your Christmas tipple like a poison and attempts to break it down in the liver. This organ can deal with large amounts of the substance, but it requires time to do so. Drink a liter of spirits in one go, and you may well die. Pace yourself, and your liver efficiency will contribute to your tolerance for seasonal celebrations. A regular drinker will have more enzymes for breaking down alcohol than the occasional tippler.

After you drink one half-pint of beer or one glass of wine, absorption from the gut into the bloodstream is usually complete within thirty to sixty minutes. The body turns alcohol to acetaldehyde first, then to acetic acid, and finally to carbon dioxide, providing seven calories of energy for every gram of alcohol consumed. Dieters should take that into account, in addition to the tendency of moderate amounts of alcohol to stimulate appetite.

Of all the products of the breakdown process, acetaldehyde causes much of the grief of a hangover. So-called drunken breath occurs when this molecule emerges in exhaling. But the impact can be much more profound. Aldehydes are linked to flushing, increased heart rate, and palpitations. Aldehydes also can poison cells, and even play a role in the development of some cancers. For those reasons, the elimination of aldehydes is crucial to the health of living cells.

Genes are heavily involved in how we respond to alcohol. The first hint of genetic predisposition came from studies of drunken mice. Some strains dozed off for hours, whole others came around soon. Still others gazed vacantly at the walls of their cage, while a few rushed madly around as if they had seen Marley's ghost (or the rodent equivalent).

Two enzymes are involved in our capacity to cope with the spirit of Christmas. One, called alcohol dehydrogenase, breaks down the poison; the other attacks its even more noxious breakdown product, acetaldehyde. People unlucky enough to inherit a powerful version of the first enzyme and a weak form of the second suffer immediate and unpleasant side effects from drinking. Those thus affected are called "fast flushers." Their faces go red, and they sweat and feel ill whenever they drink. Not surprisingly, most lay off the booze. This combination of genes is much rarer in the West than elsewhere, which may be why alcohol plays such a large part in European cultures.

Differences in alcohol tolerance are not only due to the body's ability to break it down in the liver. Last year, a Japanese team found a brain chemical that helps animals recover from the effects of alcohol. The team studied mice by loading them up with alcohol, putting them on their backs in troughs, and seeing how long it took before they managed to stagger to their feet. Mice lacking the crucial enzyme, known as Fyn tyrosine kinase, took twice as long to get up as mice with a functioning gene that produced the chemical. Fyn appears to act in the hippocampus of the brain within five minutes of alcohol's arrival there. Alcohol normally causes brain cells in the region to become less responsive and less likely to fire signals, but the enzyme helps them to recover. "The absence of the Fyn enzyme makes people more deeply drunk and delays them from becoming sober," says Hiroaki Niki at the Riken Brain Science Institute, who led the research. Individual differences in enzyme levels could explain why some people can hold their drink during a Christmas party, while others reel under its influence.

Whoa!

One of the most disturbing feelings experienced by many revelers is "whirling-pit syndrome." Having had not one but several Christmas drinks too many, and too quickly, sufferers of the syndrome will feel the room spin as they settle, or collapse, into an armchair or bed.

The disorientation and dizziness are caused by alcohol's disrupting a balance sensor in the inner ear. The sensor consists of a sac and three semicircular canals containing fluid. As we move, the corresponding fluid motion is detected by tiny hairs lining the organ. The hairs translate the movement into electrical impulses. The brain uses this information to help calculate balance. To do this it assumes a certain density of the fluid. But as alcohol diffuses into the fluid, this assumption breaks down, causing dizzying shifts in the sense of balance, even if there is no actual motion.

This also explains why the "hair of the dog" actually works. It is not the absolute density of the balancing fluid that causes disorientation, but *changes* in its density. During the period of sobering up, when alcohol diffuses out of the balancing fluid, the density changes may be too rapid for comfort, producing a distinctly unsteady feeling. A little alcohol may cut the rate of density change, restoring balance and removing the feeling of fragility — although there is more to dealing with hangovers than that.

The thirst of a hangover is thought to be due to alcohol's diuretic effect. It inhibits a hormone called vasopressin, which controls how the kidneys reabsorb water. As a result, more water passes to the bladder, so that drinkers visit the lavatory more often. If you drink around two glasses of wine, you lose about twice the amount of water from the body in the next two hours.

Ouch!

Scientists now think they know why throbbing headaches sometimes result from overenthusiastic drinking during Christmas festivities.

Headaches do not start in the brain, which has no pain receptors, but in the major blood vessels of the meninges — the membranes between the brain and the skull. For a long time it was thought that the pain was caused by the swelling of these blood vessels. This idea was overturned by techniques that enabled doctors to examine the vessels in conscious patients.

Now it seems that sensitization is a better candidate to explain the characteristic ache. Andrew Strassman, of the Beth Israel Deaconess Medical Center and Harvard Medical School in Boston, and colleagues provided direct evidence from studies of rats to suggest that nerve endings in the blood vessels become highly sensitive as a result of being exposed to chemicals in the bloodstream. The same kind of sensitization also primes the nerves to respond to movement, suggesting why, in some hangover headaches, the pain is worsened by coughing or sudden head motion.

The chemical culprits for sensitization are usually linked to inflammation and injury. But in the case of a hangover, they are due to the breakdown products of drink, notably constituents called congeners, which are formed during fermentation. Brandy, cheap dark rum, and bad red wine produce the worst hangovers, being high in congeners. By contrast, pure alcohol, gin, and vodka are relatively low in congeners, producing less severe hangovers.

Congeners may explain why the mythical "hair of the dog" can reduce the severity of a hangover. A link has been shown between the breakdown of one congener, methanol, into formaldehyde and formic acid and the onset of hangover symptoms. Ethanol blocks the formation of formaldehyde and formic acid, possibly explaining the partly restorative effect.

How to Cope

Of course, the best hangover remedy is not to drink in the first place. If you must, then eat. A full stomach retards the passage of alcohol into the small intestine, delaying the effect. That is why a cocktail on an empty stomach has a bigger kick than a couple of drinks after a heavy meal. A glass or two of milk can also help. Another suggested remedy is to take N-acetyl-cysteine, an amino acid supplement available in many health food stores that helps replenish the body with glutathione, part of its detox machinery. Exercise, however, doesn't do much to help a hangover: the rate of breakdown of alcohol stays quite steady, at around 20 grams (3/4 ounce) of whiskey per hour for a man weighing 75 kilograms (11 3/4 stone). And coffee merely acts as a stimulant to counteract some of the alcohol's depressive effects. It is also thought unlikely that traditional remedies, such as sweating in a sauna or taking cold showers, can really help. Sobering up is simply a matter of time.

Sheryl Smith of the Allegheny University of the Health Sciences, Philadelphia, linked hangovers to changes in a receptor in the brain, where a messenger chemical called GABA acts to dull nerve circuits. The changes are caused by the increased production of a protein that forms part of the receptor, called the alpha-4 subunit. When this occurs, the GABA receptor is less efficient, so it cannot calm those sizzling circuits quite so easily. A drug that blocks the production of the alpha-4 protein could moderate the symptoms of hangovers. However, a completely effective hangover treatment will probably always be beyond the reach of medical science, according to Ian Calder, a consultant anaesthetist at the National Hospital for Neurology and Neurosurgery, London: the hangover is too complex a condition to cure.

In an editorial for the *British Medical Journal,* he maintained that ethanol (alcohol) was only partly responsible for the all-too-familiar symptoms of thirst, headache, fatigue, nausea, sweating, tremor, and anxiety. Other factors — such as lack of sleep, smoking, overeating, snoring, and any unusual physical or emotional incidents that occur during a drunken night out — also play a part, making a truly effective treatment for hangovers unattainable. Moreover, he believes that a completely effective remedy is "arguably undesirable," since the fear of hangover prompts most people to moderate their alcohol intake. Even moderate amounts can be damaging, so a painful penalty the morning after is in our best interest.

WHAT HAPPENED?

If you *think* you enjoyed the seasonal celebrations, but the actual details are rather hazy, the reason for your forgetfulness has emerged from a study of the protein c-Fos, which provides a general measure of the activity of brain cells. It turns out that alcohol affects our memory for details more than our memory of major events. The study, described in the journal *Molecular Psychiatry,* followed up earlier research suggesting that one disturbing effect of alcohol on cognitive functions was due to its effects on the hippocampus, the brain structure responsible for the formation of complex memories and orientation in time and space.

In a more recent study, scientists from the Scripps Research Institute in La Jolla, California, traced the levels of the protein c-Fos in rats that had gulped down the equivalent of three or four drinks for humans. The team found that the level of this protein was stimulated in several regions of the brain, particularly those involved in regulating emotions and behavioral motivation and in processing sensory stimulation. It was selectively decreased only in the hippocampus.

Significantly, this moderate dose of alcohol did not only decrease the levels of c-Fos in the hippocampus of otherwise untreated rats, but it also blocked the increased activity that typically occurs when the animal is exposed to a novel environment. This may explain alcohol's effect in suppressing the ability to remember new information. A lower dose of alcohol (corresponding to one to two drinks) also decreased the levels of c-Fos in otherwise untreated rats, but it did not block the response to a novel environment, suggesting that this lower dose is not enough to disrupt the processing of new information.

These studies help to explain why after recovering from a binge, one may not remember specific experiences, such as dancing on a table, or much about the place where drinking occurred. However, you may still have a feeling that a good (or bad) time was had.

The Alcohol Debate

Alcohol, in small quantities, is good for you. It can ward off heart attacks and cardiovascular disease and stimulate digestion. Marie Antoinette believed that wine could even help prevent wrinkles. But boost the daily dose of this drug beyond four units for a man and three for a woman, and these health benefits decline. Beyond that point, alcohol raises blood pressure, actually adding to the risk of cardiovascular disease; it can impair cognitive function, contributing to accidents; and it increases the risk of certain cancers and liver disease.

This pattern is summed up by a U-shaped relationship between death rate and alcohol intake. This *U* has been traced out by various studies, for instance by the American Cancer Society in Atlanta and a thirteen-year survey of twelve thousand male doctors published in the *British Medical Journal.* The lowest death rates were found among doctors who drank up to twenty-one units per week — the equivalent of one and a half pints of beer a day.

Alcohol in moderation is thought to boost blood levels of high-density lipoproteins, or "good" cholesterol, which are associated with a lower risk of cardiovascular disease. Alcohol also might prevent clots from forming in arteries. One major study was published in 1997 by Patrick McElduff and Annette Dobson of the University of Newcastle, New South Wales, as part of a World Health Organization project. They concluded from an investigation of around twelve thousand heart attacks (one-quarter of which were fatal) that the risk is lowest among people who consume between one and two drinks daily on five or six days a week.

Although such studies have indicated that moderate drinking decreases the likelihood of a person's suffering a heart attack in the first place, there may even be potential benefits to drinking *after* a heart attack. In the *Proceedings of the National Academy of Sciences,* Vincent Figueredo and colleagues at the University of California at San Francisco reported that they had found a mechanism by which moderate drinking may reduce the damage caused by a heart attack, thereby improving chances of long-term survival.

The type of alcohol may be a factor when it comes to its beneficial effects. In research to determine the relationship between different types of alcoholic beverages and "oxidative stress" — cell damage caused by highly reactive chemical intermediates called free radicals, which play a role in many chronic diseases — a University of New York at Buffalo study has shown that people who drink wine appear to experience less alcohol-related oxidative stress than people who drink beer or liquor. "The difference was small but significant," according to lead author Maurizio Trevisan.

Geoff Lowe of the University of Hull, England, believes that psychological factors, notably laughter, also contribute to the benefits of alcohol. By hanging around in local pubs, he observed what every drinker knows — that a pint or two makes people jollier.

In a survey of 332 social drinkers, humor and laughter were found to be more prevalent in the everyday lives of heavy drinkers. A second experiment showed that people drinking alcohol laughed more than those consuming nonalcoholic drinks while watching the comedy film *Naked Gun*. In a third study observers monitored how much young people laughed in bars and pubs. Laughter was again significantly higher for groups of alcohol drinkers.

"Taken together with evidence that laughter can serve as a stress moderator and enhance immune function," Lowe says, "our observations suggest that the positive influence of moderate alcohol consumption on health and longevity may at least partly be due to its mood-enhancing and stress-buffering properties."

Alcohol also acts as a ready-made source of energy, with at least 70 percent of the calories it provides immediately available for use. A single shot of gin or vodka contains about 125 calories, and a liter of Guinness around 370 calories, for instance. Intriguingly, Colorado State University researchers found no link between moderate red wine consumption — that is, in amounts less than 5 percent of an individual's daily calories — and weight gain. A possible explanation is that calories from alcohol may be metabolized differently than other food calories.

But what does all this mean in terms of the effects of alcohol on life expectancy? A 1991 study in the journal *Circulation* calculated that if all heart disease were eliminated, the average life expectancy of a thirty-five-year-old would increase by about three years. If moderate alcohol consumption confers a net health benefit, some portion of those three added years of life, perhaps a few months, might be attributed to it.

But that is a statistical average, and every individual is different. The same does not apply to younger drinkers, among whom alcohol is thought to cause more deaths by increasing the risk of accidents and brawls in bars than it might prevent by reducing heart disease.

As we have already seen, there is also an array of genetic factors that affect our sensitivity to drinking. One day it may be possible to take account of how each of us individually responds to alcohol. DNA analysis chips that can screen an individual's genetic makeup are already on the market. Until now the primary application of these devices has been to detect the risk of inherited diseases. In a few years one may be able to purchase the ultimate stocking stuffer: a DNA chip that allows you to check your own genetic makeup and to assess how you cope with your own favorite poison.

Until then we all have to take carefully into account the downside of alcohol. First, those calories we do obtain from alcohol are nutritionally empty, and at high levels of consumption, alcohol can inhibit the absorption of both fat and vitamins such as thiamine, riboflavin, niacin, and folic acid from the intestine. It also reduces the body's levels of magnesium, calcium, and zinc. Chronic alcoholism is often associated with malnutrition.

When drinking exceeds moderate levels — more than about three units a day — it starts to increase a person's susceptibility to cardiovascular disease, for which hypertension, or high blood pressure, is a well-known risk factor. Alcohol is second only to obesity in its contribution to hypertension.

Besides liver disease, alcohol also raises one's susceptibility to certain types of stroke and to cancers of the mouth, throat, and liver. One paper in *Nature Medicine* showed that rats injected with cancer cells developed up to ten times more tumors if first given a large dose of alcohol — equivalent to an alcoholic binge — than if they abstained. The researchers, including Shamgar Ben-Eliyahu of Tel Aviv University, concluded that alcohol suppresses the activity of the immune system's natural killer cells, which are known to play a crucial role in ridding the body of tumor cells.

High-Tech Wines

Within a decade or so, it will be possible to enjoy more wine at Christmas, without the risk of a hangover, thanks to advances in the development of genetically engineered grapevines. The competition to produce these biotech wines is intense, with rival teams at work in the United States, France, Israel, and Australia.

Nigel Scott, chief research scientist of the CSIRO Division of Horticulture in Adelaide, is head of one effort to introduce new genes into grapes. For some traits, such as resistance to pests and spoilage, this is relatively easy. For other purposes, such as altering flavor to ensure a spicy Christmas glühwein, successful gene tinkering is a much more difficult prospect.

At the CSIRO's Adelaide laboratory, Mark Thomas and Tricia Franks have studied techniques to introduce new genes into sultana grapevines, paying particular attention to the plant hormones that control regeneration from a few cells to embryo to grapevine. Working with the Cooperative Research Center for Viticulture, the Grape and Wine Research and Development Corporation, and Dried Fruits Research and Development Corporation, they have been successful in introducing genes into cells and then, most important, growing a grapevine from them.

For reasons that are not understood, the parts of the plant that respond best to genetic engineering are filaments taken from the stalk that ends in the plant's anther, the lobes in which pollen matures. Using *Agrobacterium tumefaciens,* a bacterium that naturally transmits genes into plant roots, a marker gene called Gus was introduced into cells from the sultana filaments. The engineered cells were then cultured in the laboratory into embryos that could be grown into full-size plants.

Now the team wants to switch off the gene responsible for polyphenoloxidase, an enzyme that causes sultanas to go brown, so that they are longer lasting as a result. "We believe we will soon be able to turn off this gene and produce superb golden-colored sultanas," Scott says.

The hunt also is on for other genes to alter. Simon Robinson and Ian Dry, who found the polyphenoloxidase gene, are now searching for grape genes that aid defense against fungal diseases. Boosting the expression, or use, of these genes in grapevines could provide improved resistance. Expression of genes involved in the ripening of grapes — for instance, genes involved in sugar accumulation, berry softening, and synthesis of flavor compounds — also is being studied. The team hopes to extend the work to other grape varieties as well. "We want to try introducing genes into chardonnay, pinot, and Shiraz," says Scott.

Another aim — one close to the heart of the Christmas tippler — is to produce a grape with reduced sugar content and thus reduced alcohol. The wine would be superior in taste to the low-alcohol wines now on the market. Then Santa can have two bottles of wine with his Christmas lunch and still be able to drive his sleigh in a straight line.

Fatlash

A great deal of scientific endeavor has focused on the downside of alcohol. The start of this decade saw the establishment of a scientific association for the study of enjoyment. Specialists in fields as diverse as psychology, pharmacology, and neurochemistry united in a common purpose: to challenge the hectoring of the health education lobby by demonstrating that many of the pleasures it frowns on during the Christmas festivities — from eating fatty foods to drinking rich red wines — are good in moderation and that feeling guilty about them is bad for you.

During the late 1980s David Warburton of Reading University in England became increasingly concerned that, though abundant research was being conducted on anxiety and stress, few studies were being made of pleasure. He was alarmed, for instance, by reports that parents in the grip of a high-fiber fad were putting toddlers on low-fat diets, depriving them of the cholesterol they need to develop their brains and sex hormones.

He set up ARISE — Associates for Research into the Science of Enjoyment — to explore the idea that pleasure is an essential part of life but that experiencing guilt about it can lead to health problems. "When people are brooding over their guilt, they become more absentminded and error-prone," he says. "Chronic guilt can induce stress and depression, leading to eating disorders. It can contribute to infections, ulcers, heart problems, and even brain damage."

With enthusiastic backing from food manufacturers, he searched for ARISE recruits around the world. In California, the world capital of health faddism, he found Faith

Fitzgerald, a physician at the Davis Medical Center, who believes that many doctors want to exert "tyrannical" control over people's behavior. In Dublin, James McCormick, a community health physician at Trinity College, came forward to declare that it is better to encourage people to lead lives of modified hedonism, so that they might enjoy them to the full, than to portray life as a perilous journey beset with avoidable dangers. Italian psychiatrist Vittorino Andreoli is interested in how a society of guilt can be transformed into one of pleasure. He located scientists in Germany and Australia to examine how moderate alcohol consumption can be healthy.

The broader conclusion to be drawn from a variety of research, according to Warburton, is that pleasure is itself an antidote to the harmful stresses of modern life. Even an enjoyable film or record was found to raise levels of the antibody immunoglobulin A, indicating a strengthening of the immune system. By contrast, the kind of stress associated with guilt had the opposite effect. It made people more prone to disease by lowering the production of lymphocytes, the white blood cells that fight infection. Indeed, people who are tired, depressed, or bullied at work are more likely than happier colleagues to come down with colds and flu. ARISE surveys suggest that nearly four out of ten people would enjoy many of their basic pleasures far more if they did not feel so guilty about them.

"This medical evidence that pleasure is good for you is a useful riposte to the moralistic self-righteousness of those who believe that there is only one way to live your life — theirs," Warburton says. So long as you do not go overboard, the indulgence that goes with seasonal good cheer may be good for you. I, for one, will drink to that.

10

CHRISTMAS BLUES AND SEASONAL MOODS

He was checked in his transports by the churches ringing out the lustiest peals he had ever heard. Clash, clang, hammer; ding, dong, bell. Bell, dong, ding; hammer, clang, clash! Oh, glorious, glorious!

Running to the window, he opened it and put out his head. No fog, no mist; clear, bright jovial, stirring, cold; cold, piping for the blood to dance to; Golden sunlight; Heavenly sky; sweet fresh air; merry bells. Oh, glorious! Glorious! . . . Christmas Day!

CHARLES DICKENS, A CHRISTMAS CAROL

SMILING, HAPPY PEOPLE HAVE LONG BEEN A common feature of Christmas cards, following a tradition that suggests that human behavior is somehow transformed at this time of year. Take Dickens, for example. From his first account of the festivities in *Sketches by Boz* to the beatification of Scrooge in *A Christmas Carol,* he emphasized again and again that Christmas is a time for social altruism, warmth, and friendship.

Is there any evidence that people act differently in any quantifiable way around the holidays? Well, surprisingly, there is. Some of these influences are positive, others negative, and a few downright odd. All go beyond obvious manifestations of Christmas spirit such as hangovers, family gatherings, and expanding waistlines.

The short winter days can indeed make some of us sink into gloom and go on carbohydrate binges. But there is plenty of good news. The widely held view that the suicide rate soars during December is undermined by sober mortality statistics. In fact, for the moribund a celebration such as Christmas, Hanukkah, or Kwanzaa can actually help delay the day of reckoning. For those with a religious belief, there is tantalizing evidence of a "faith factor" to help fight disease. And those born on Christmas Day may well get a better start in life, particularly if they want to become high-ranking clergy.

For all of us Christmas may be in some way connected with a peak in the levels of a signaling chemical — a pheromone that triggers a subconscious response in the human brain. Before we tackle this malodorous manifestation of Christmas spirit, let's start with the bad news.

Christmas Gloom

Does Christmas drive you crazy? There is a widespread belief that holidays and festivals can increase the incidence and expression of psychiatric disorders. The Swiss even have a name for it, *Weihnachtscholer*, or "Christmas unhappiness." It's one of the great Yuletide ironies: in a season celebrating joy, fellowship, and charity, the holiday blues may drive the suicide rate up.

Actually, recent developments have cast doubt on this notion, but there is still a suspicion that some people experience the holidays as the season of bad cheer and misery. Perhaps a slackening of the prohibitions against self-indulgence at Christmas can trigger disease. Or maybe the seeds of misery are sown when hopes are raised by magical wish fulfillment, then dashed after the holiday by the frustrations of everyday life.

Some of the theories that have been proposed to explain this gloomy side effect of the Yuletide season seem quite bizarre. One, put forward in the summer of 1955 by psychiatrist Bryce Boyer in the *Journal of the American Psychoanalytic Association,* attributed the depression to "reawakened conflicts related to unresolved sibling rivalries. . . . It is suggested tentatively that partly because the holiday celebrates the birth of a Child so favoured that competition with Him is futile, earlier memories, especially of oral frustration, are rekindled."

Boyer based his ideas on a sample of seventeen patients, one example of whom was "Mrs. W," a thirty-year-old childless housewife and "inactive Episcopalian" who would annually become depressed and preoccupied with religious thoughts by the middle of December. She had always been convinced that she was unwanted and had blamed this rejection on being female. According to Boyer, her ruminations revolved around one theme: "If only I would turn to religion, God would give me a penis."

Boyer concluded that his patients "uniformly" sought to obtain penises, "with which they imagined they could woo their mothers to give them the love which they felt had been previously showered upon their siblings. . . . They at times identified with Christ in an attempt to deny their own inferiority and to obtain the favouritism which would be His just due."

Equally extraordinary is the attempt in 1944 by the psychologist Richard Sterba to compare behavior around Christmas to the customs surrounding childbirth: the preparatory excitement, secret anticipation, last-minute flurry of preparation, prohibition about entering the rooms containing gifts, and relief of tension afforded by the delivery of a gift, whether a baby or a sweater knitted by an aunt. "It is not surprising that the presents come down the chimney since the fireplace and chimney signify vulva and vagina in the unconscious and the child-present thus comes out of the birth canal. This casts some light on the figure of Santa Claus. He, no doubt, is a father representative." This, Sterba concluded, is why susceptible people develop mental illness around Christmas: it stirs up unconscious fantasies and unresolved conflicts about childbirth.

Another article, presented in 1954 to the American Psychoanalytic Association in St. Louis, put forward the concept of a holiday syndrome, lasting from Thanksgiving until after New Year's Day, which was characterized by depression, "diffuse anxiety," nostalgia, and "wishes for magical resolution of problems." This syndrome reached a zenith at Christmas, probably because the sufferers had difficulty establishing close emotional ties and, as a result, felt isolated, lonely, and bored.

Certainly the media have annually presented plenty of articles about Christmas blues and expended acres of newsprint on how to deal with them. But is the perceived peak in suicides a real phenomenon? No, according to several strands of research. In Britain one study of 22,169 attempted suicides over nineteen years revealed that the rate among women dropped to three-quarters the usual level during the Christmas period. Similarly, there was no evidence of a monthly or seasonal variation among men.

Comparable results have emerged from studies in the United States, again suggesting that the syndrome, if it exists, does not produce a strong enough effect to show up in monthly statistics on suicide. Research carried out in the 1960s and 1970s found that the numbers of suicides in December and January were average or low. One investigation, conducted in North Carolina, was published in the *Archives of General Psychiatry* in 1974. William Zung and Richard Green studied 3,672 suicides and 3,258 admissions to the Durham Veterans Administration Hospital Psychiatry Service. They found no statistically significant increase in psychiatric admissions to hospitals or in suicides at Christmas or during any other holiday period.

Overall, American statistics compiled over the past two decades show that December records the fewest suicides of any month. Suicide rates tend to rise in spring and summer and edge down in fall and winter. "What you find is not an increase but a drop in suicide during the holidays," says David Phillips, a sociologist who has studied these seasonal trends. In 1991, for example, the months of April and June tied for the highest average daily suicide count, followed in order by August, July, May, and September. December ranked last. "The holidays appear to be providing some psychological and social protection against suicide, but the nature of this protection is currently unknown," he concluded after analyzing the timing of more than 180,000 suicides from the late 1970s.

The festive celebrations during the dark days of winter may have one unexpected bonus for those who are near death. There is growing evidence that the moribund can successfully "bargain with God" or exercise "willpower" to manage to live until some important occasion. Phillips, working with Elliot King, set out to determine whether a dying person can indeed strike a deal with his maker to survive for a few extra days to join in a celebration such as Christmas, Hanukkah, or Kwanzaa.

They decided to test the idea by comparing deaths among Jews and non-Jews before and after the Jewish holiday Passover. This was an attractive holiday to study because its actual date shifts around the calendar by about four weeks a year, allowing separation of holiday effects from purely seasonal effects, notably the change in the hours of daylight.

The most important event of Passover is a ritual dinner called the Seder, which celebrates the Exodus of the Israelites from Egypt, their receiving the Torah, and their entry into the land of Canaan. More than three-quarters of American Jews attend a Passover Seder.

Although California death certificates do not list the faith of the deceased, Phillips and King described in the journal *Lancet* how they sidestepped this difficulty to establish their sample population. They focused on 1,919 death certificates of men who had characteristically Jewish names, such as Cohen, during the weeks before and after Passover from 1966 to 1984. For controls they used a list of surnames, such as Rose and Green, that are common in both Jewish and non-Jewish populations. For other controls the researchers isolated a group of people coded as Japanese or Chinese on death certificates.

The study did indeed reveal a dip in deaths during the week before Passover and an 8 percent increase during the week after. Among 625 men with "unambiguously" Jewish names, they found a 25 percent increase in the week after Passover. When it came to weekend Passovers, when family gatherings are likely to be larger, there were 61 percent more deaths in the week after than in the week before. The researchers found no such differences among Japanese or Chinese people.

This finding, which the authors called the "Passover effect," applied solely to adult men and was absent in adult women and in children under the age of four. The effect appeared in all three leading categories of the cause of death (heart disease, malignant tumors, and cerebrovascular disease), but not in deaths from other natural causes or from external causes.

Why did the Angel of Death "pass over" the Jewish households until after the holiday? The results, Phillips and King argued, were "consistent with two hypotheses: that the 'will to live' is associated with reduced mortality, and that communal social events can have a beneficial impact on the course of a disease."

This is by no means a modern phenomenon. The authors pointed out that both John Adams, who was born in 1735, and Thomas Jefferson, who was born in 1743, managed to live until July 4, 1826, the fiftieth anniversary of the signing of the Declaration of Independence. According to Jefferson's physician, his last words were, "Is it the Fourth?"

Not everyone was convinced that the key to the Passover effect lay in the urge to live to see the big day, rather than something that happened to individuals during the festivities themselves. Michael Baum of King's College School of Medicine, London, commented in *Lancet* that the statistics may be explained by diet. "The Passover diet among observant Jews has to be free of all leavening and thus is highly costive," he explained. A diet rich in matzo (unleavened bread) contains little bran and has a constipating effect, which results in "straining at stool," raising blood pressure and increasing the risk of heart attack or stroke.

Baum raised another issue concerning women. Deaths of Jewish women were *not* postponed until after the Passover, which the California team argued was because they take a lesser part in the ceremony. But Baum countered that Passover drives Jewish women nearly to breakdown: "Large numbers of wives from the ultra-Orthodox community would crack up under the psychological strain of preparation for the Passover, seeking refuge in the hospital ward until the Angel of Death passed over the household and normal services could be resumed."

Undeterred, the Californians have found still more evidence of the death-defying effects of seasonal celebrations. For the past twenty-five years, in the week immediately before the traditional Chinese Harvest Moon Festival, they have observed more evidence of the strange phenomenon in the local Chinese community. They described their findings in 1990 in the *Journal of the American Medical Association.*

The death rate for older women, who play a central role in the celebration, suddenly dips by as much as 35 percent, as if those who would have died postponed their passing to be part of the occasion. Then, after a corresponding increase in the death rate immediately after the festival, the rate returns to its usual level. These observations seem to mirror the earlier finding among Jewish men at Passover. A control group, drawn from people for whom the Chinese holiday had no symbolic significance, showed no such dip in mortality.

This phenomenon cannot be explained by conventional hypotheses, says Phillips. The Chinese women, for example, are not dying in greater numbers after the holiday because of extra stress or overeating. Nor can the behavior of either the Chinese women or the Jewish men be explained by any fixed monthly pattern in mortality, because both holidays move around the calendar from year to year.

There may in fact be a "causal pathway" linking psychological and biological events, and this phenomenon may be far broader than originally recognized. Personal events of great symbolic importance — a fiftieth wedding anniversary, for example — also may have demonstrable short-term effects on mortality. "The best available explanation is that deaths of some people are postponed until they have reached an occasion that is important to them," Phillips says. The moribund can put off death for these "symbolically meaningful" occasions, such as Christmas.

Why We Are SAD

The decline in suicide rates during the holidays does not necessarily mean that depression also declines. It is conceivable that although fewer people actually carry out the threat to kill themselves, the number of people who come to a hospital emergency room or a doctor's office complaining of melancholy actually rises during the holidays.

For years the winter blues were written off as psychiatric curiosities. But in the past few years, the medical community has acknowledged that these symptoms represent a real illness. There is undoubtedly a propensity for some individuals to feel sad and gloomy as a result of seasonal affective disorder (SAD), a syndrome characterized by recurrent depressions that occur at the same time annually in response to the decline in daylight hours. As a result, for a significant fraction of the population that lives at high latitudes, winter days can mean misery.

Norman Rosenthal, a psychiatrist with the U.S. National Institute of Mental Health, first coined the term in 1984 to describe how short days and reduced sunlight can trigger a change in brain chemistry. Winter after winter, certain people experience lethargy and fatigue, sadness and despair. SAD can disrupt personal relationships and cause its victims to overeat, gain weight, and become indifferent toward their jobs.

It is now recognized that women are four times more likely to be affected by SAD than men and that the disorder in women usually starts after puberty and diminishes after menopause. Rosenthal speculates that female reproductive hormones somehow sensitize the brain to the effects of light deprivation.

The incidence of SAD also is directly related to latitude, varying according to the day-night cycle, with those farthest from the equator and closer to a white Christmas most often affected. Studies have shown SAD symptoms in about 10 percent of people in New Hampshire, which has long winter nights, but only about 1.5 percent of people are affected in Florida, where the winter days are longer.

The good news is that these dark moods can be banished to some extent by exposure to intense light. The reason this treatment works has been a puzzle, because the light can be applied during the day, leaving the effective short day length of winter unchanged. Recently, however, a milestone experiment on hamsters by a team at Northwestern University has shed new light on the mystery of how illumination can adjust brain chemistry to make Christmas more tolerable for SAD sufferers.

For a number of years it has been known that a brain-signaling chemical called serotonin is somehow involved in depression. Fred Turek and his colleagues at Northwestern found that pulses of light administered in the middle of the day altered the way hamsters responded to the effects of serotonin on the biological clock. These results indicate that light can alter the way neurons in the brain respond to serotonin. SAD may not simply be caused by a lack of light but by the effect of light on serotonin metabolism.

Christmas Excitement

But why celebrate at all this time of year? One explanation dates back to prehistoric fascination with the sun. Another may have been uncovered by smell researchers, who believe that chemicals in human sweat exert an effect on the brain over and above that of religion, excessive consumption of food and drink, and Yuletide bonhomie. The chemicals are called semiochemicals, or pheromones. It is only recently that scientists have provided convincing evidence that these airborne chemical signals can affect the human body, with the discovery this year by researchers at the University of Chicago that armpit secretions in women cause menstrual changes in other women. The physical and mental reactions triggered in us when we smell sweat are still not fully understood, but they may have a bearing on seasonal festivities. A study conducted at the Monell Chemical Senses Center in Philadelphia suggests that every December two chemicals found in male sweat, androsterone and androstenol, rise to their highest levels. "The peak is right at the end of December," notes David Kelly of the chemistry department at the

University of Wales at Cardiff. "I have speculated that the cause of Christmas may be a sudden increase in the production of this material," says Kelly.

These chemicals are not actually responsible for the particular smell of sweat but are nonetheless musky-smelling steroid molecules. Somewhat out of keeping with the spirit of the season, they are both found in the testes of pigs. Androstenone, a closely related compound, can be bought in spray form by pig farmers who need an aphrodisiac to put their porcine charges in a romantic mood.

Semiochemicals influence behavior in subtle ways. Androstenone, a closely related compound, does not look particularly promising as a human sex attractant, since some find its smell offputting. However, when one of several chairs in a doctor's waiting room was sprayed with the steroid, it was found that women preferred to sit on that chair. Another experiment showed a similar effect when the steroid was sprayed on seats in a theater.

In the female anatomy, breasts exude the most androstenone, and one explanation for its mood-altering effect is that we link its aroma with being breast-fed and thus with bonding. Kelly argues that production of these steroids may be a physiological reason many people feel more excited at this time of year. No wonder December has seen feasting since ancient times.

Seasonal Jolliness Is Good for You

Let's assume that most people are jollier and celebrate more during the holiday season. Does this heightened mood have any health implications? Studies at the Institute of Heart-

Math in Boulder Creek, California, published in 1995 in the *American Journal of Cardiology,* suggest that although anger can put the heart at risk, "sustained positive feelings" of the kind (presumably) in good supply over the holiday period can help protect against heart attacks and high blood pressure. A relaxation technique designed to manage mental and emotional stress has been shown to affect the autonomic nervous system, which controls heart rate and breathing. Alan Watkins, a doctor at Southampton General Hospital in England, says that Christmas spirit may have a "similar although somewhat smaller beneficial effect."

Although I admit this is stretching things somewhat, a number of studies on mice seem to suggest that pleasurable experiences can enrich the brain. When mice are reared in big cages and pampered with toys, comfortable nesting material, and tasty snacks — features of what one might imagine to be a traditional rodent Christmas — they are found to have more nerve cells in the hippocampus, the region of the brain linked to memory, than their more deprived brothers and sisters.

This kind of effect has been under investigation for the past four decades. A number of studies have shown that placing laboratory animals in a stimulating environment boosts performance on standard tests, such as the ability to negotiate mazes. But the biological mechanisms underlying these improvements have been hotly debated, with some scientists believing that improved performance can be completely accounted for by mechanisms of increasing and strengthening the *connections* between nerve cells in the brain.

To investigate these details, a study of the effect of environment on the brain was conducted at the Salk Institute in La Jolla, California. The team was led by Fred Gage, who believed it might be possible that the *number* of nerve cells also is important.

Gage's colleagues Gerd Kempermann and Georg Kuhn separated twenty-one-day-old mice into two groups: one housed in "standard" laboratory conditions, the other housed in a large cage "enriched" with tunnels, toys, an exercise wheel, and food treats such as apples, popcorn, and whole-grain nibble bars. (For the purposes of this book, we will consider this the equivalent of comparing a group that has a boring time over Christmas with another that gets rather carried away with the celebrations.)

After forty days, the two groups of mice were compared in a water maze test. As expected, the group raised in the enriched conditions performed significantly better than the control group. The scientists also found that the hippocampus of enriched mice contained an average of forty thousand more nerve cells than those of the controls.

"We were overwhelmed by the magnitude of the increase, which represents a gain of 15 percent in the number of these nerve cells," Gage says. He believes that the mice in the enriched group were not necessarily generating more brain cells, since those cells seemed to be dividing at the same rate in both groups. Rather, an enriched environment appears to have fostered the *survival* of new brain cells.

Several days before the experiment's end, when the mice were around sixty days old, some of them were injected with a dye that made new brain cells fluorescent and thus identifiable under the microscope. The survival rate of newborn cells in the enriched animals was 60 percent higher than in the control group. The results are remarkable because the experiments were not carried out in infant mice, whose brains one would expect to be more plastic, but rather in older animals. If the human brain works like that of the mice, older people also can improve the "architecture" of their brains by enriching their lifestyles — which they can do quite easily with seasonal celebrations.

Christmas Birthdays and BIRG

The beneficial effects of Christmas don't stop there. Those who are born on Christmas Day may well have a better start in life, thanks to a phenomenon called BIRG.

Psychologists have noted that after a victory of their school football team, college students were more likely to wear school sweatshirts than they were following a team defeat. In addition, following a win students would say, "*We* won." But following a loss they would mutter, "*They* lost." This tendency to associate with success is called "basking in reflected glory," or BIRG.

A research team from the University of California at Davis and the Israel Institute of Technology in Haifa has sought another path to reflected glory by investigating the relationship between the dates of national holidays, notably Christmas, and the reported dates of well-known individuals' births. For example, one of the most influential scientists of all, Sir Isaac Newton, was born on Christmas Day in 1642.

The researchers found that a disproportionate number of famous people who were born during seven-day periods centering on Christmas Day were born on the holiday itself. This effect was stronger when it came to Christmas compared with the Fourth of July and New Year's Day. And the link was demonstrable whether it involved those listed in *Who's Who* or members of the U.S. Congress.

According to the BIRG interpretation, the association of a birthday with "positively evaluated stimuli" — a national holiday — enhances a person's image. Perhaps the parents of a child born on Christmas Day believe that he or she is special, like baby Jesus, and will have an edge in life as a result. Perhaps the child holds this conviction, so that once this bond between holiday and birthday is forged, it boosts confidence and becomes a self-fulfilling prophecy.

The team that made the find — Albert Harrison, Nancy Struthers, and Michael Moore — decided to take this research one step further and study whether people born around Christmas were exceptionally likely to become religious. Although the finding was tentative, it did provide intriguing support for the BIRG effect. In comparison to low-ranking Christian clergy, high-ranking clergy were more likely to have been born on December 25. The team concluded that being born on Christmas Day is associated with upward mobility within Christian church hierarchies.

The Faith Factor

In 1872 the progenitor of human genetics, Francis Galton, said in his *Statistical Inquiries into the Efficacy of Prayer* that he could find no evidence that prayer is effective. By that he meant he found no scientific grounds for believing that prayers are answered. Yet he conceded that prayer can strengthen resolve and relieve distress. Today there does indeed seem to be evidence that those who respect religious traditions, and thus take the seasonal festivities seriously, can expect a healthier life.

Studies of this so-called faith factor face a number of

methodological difficulties. To some "religion" means innate religiosity; to others it means church attendance or orthodoxy. Add to that the problem of disentangling the interplay between religion and mental health, the effects of religion on lifestyle, and how a decline in physical health may make someone turn to religion as a durable source of hope, meaning, and purpose, and you can see that this field is fraught with problems.

Nonetheless, a number of major studies consistently point in the same direction. One study of four thousand randomly selected people in North Carolina suggests that older people who attend religious services are both less depressed and physically healthier than those who do not.

Funded by the U.S. National Institute on Aging, the study is the largest of its kind, and its lead author, Harold Koenig, has concluded that "church-related activity may prevent illness both by a direct effect, using prayer or Scripture reading as coping behaviors, as well as by an indirect effect through its influence on health behaviors." He has outlined his ideas in the book *Is Religion Good for Your Health?*

Koenig first became interested in this subject after noting anecdotal evidence of a faith factor. "I was seeing patient after patient using religion in some way to help them cope," he explains, citing activities such as praying, saying the rosary, and reading the Bible. "I got intrigued by whether it was really helping them." In nearly a dozen studies in settings ranging from prisons to hospitals to individuals' own homes, he discovered that the answer is yes. Elderly religious people are also less likely to be depressed or anxious.

Some of the influence of religion is obvious. Praying and other religious coping strategies can help older people moderate harmful stress. Religious senior citizens may be more compliant and thus more likely to adhere to regimens prescribed by their doctors. In addition, older churchgoers generally have an entire congregation watching out for them. The devout also are less likely to indulge in risky sexual activities. "Active religious participation may indirectly prevent health problems due to poor diet, substance abuse, smoking, self-destructive behaviors, or unsafe sexual practices, because these activities are discouraged by most religious groups," Koenig notes.

But are there more direct effects of religion on health? Many scientists suspect that there is a connection between state of mind and the health of the body. Even if you accept that premise, however, a "chicken or egg" problem arises: does an optimistic outlook promote a healthy body or vice versa?

Results of a study of one thousand veterans to investigate the relationship between religion and mental health, published in the *American Journal of Psychiatry,* show that the degree of religious belief correlates with *future* depression or mental well-being. Efforts to reproduce these results in nonveterans at Duke University Medical Center "look very promising," Koenig says.

This year, a related study of eighty-seven depressed patients hospitalized for medical conditions like heart disease and stroke, showed that those who scored high in "intrinsic religiosity," as measured by a scientifically validated questionnaire, recovered faster than other patients. "This is the first study to show that religious faith by itself, independent of medical intervention and quality of life issues, can help older people recover from a serious mental disorder," says Koenig. "What we may be able to do someday is to predict — from the strength of belief and change in symptoms — who is going to get better and who is going to get worse in terms of depression and so on."

Religion's salutary effect isn't limited to mental health, Koenig adds. Preliminary data suggest that elderly people who are religious have lower blood pressure and lower death rates from coronary heart disease. Greater religiousness has been associated with fewer strokes, lower rates of death from coronary artery disease, lower mortality following cardiac surgery, and longer survival in general.

You do not have to invoke the helpful hand of God to explain this. Because religious beliefs and behaviors may help a person cope with stress, the negative effects of stress on the body may be reduced or avoided. Depression, anxiety, or psychological turmoil results in the release from the adrenal glands of the stress hormones cortisol and epinephrine. These chemicals prepare the body to either confront or escape the danger. (This is called the "fight or flight" response.) When such stress is prolonged over weeks or months, the release of these substances can adversely affect the body's immune and cardiovascular systems, increasing the risk of disease.

"Religious beliefs or practices may play a role in preventing these harmful neuroendocrine responses," Koenig explains. "Some of the first evidence for a connection between religious involvement and immune system function has recently been discovered in a sample of seventeen hundred subjects in a study at the Duke University Medical Center, where low church attendance has been associated with higher levels if interleukin-6, an inflammatory cytokine indicative of immune system dysregulation."

Other studies seem to complement the idea that religion taps this mind-body link. Church or synagogue attendance has been associated with greater social contact, which in turn may be associated with better mental and physical health. This has been shown to be a significant factor when it comes to susceptibility to colds, for example.

People with more types of social ties are less susceptible to colds, produce less mucus, are able to clear their nasal passages more effectively, and release less virus than people with fewer social ties, according to a study by Sheldon Cohen of Carnegie Mellon University, published in the *Journal of the American Medical Association.* He and his colleagues exposed 276 healthy volunteers, ages eighteen to fifty-five years, to cold viruses to examine the association between cold virus resistance and diverse ties to family, friends, work, and community. Their findings suggest, counter to expectations, that the more people you meet at Christmas, whether in church or at a party, the less likely you are to get a cold. They theorize that participation in a more diverse social network may influence the motivation to care for oneself by promoting feelings of self-worth, responsibility, control, and meaning in life. Greater network diversity also has been related to less anxiety, depression, and "nonspecific psychological distress."

Cohen and colleagues also cite information from other studies showing that people with multiple social ties live longer than those who are members of fewer social groups. In a similar way, religious belief can help people cope with a feeling of hopelessness, which has been linked to hardened arteries by Susan Everson of the Public Health Institute in Berkeley, California. Her four-year study of 942 middle-aged men, reported in 1997 in the journal *Arteriosclerosis, Thrombosis and Vascular Biology,* links this state of mind — defined as feeling like a failure or having an uncertain future — to a faster progression of arteriosclerosis, a progressive disease in which fat, cholesterol, cellular waste products, and calcium collect in the blood vessels, impairing their ability to deliver oxygen and nutrients to the body and setting the stage for a heart attack or stroke.

Everson says that those who reported high levels of hopelessness after four years had a 20 percent greater increase in arteriosclerosis than those with lower levels of hopelessness. "This is the same magnitude of increased risk that one sees in comparing a pack-a-day smoker to a nonsmoker," she comments. The study gives more support to the "long-held belief that giving up hope has adverse physical and mental health consequences."

That, of course, begs another question: how far would the health benefits of a religious belief extend to other activities that also involve a lot of social networking or some kind of lifestyle philosophy — for instance, yoga or transcendental meditation (TM) classes? "They have not been compared in any head-to-head manner that I am aware of," Koenig remarks. "My sense is that yoga, TM classes, or social networking, in the absence of a belief system that stresses life commitment and accountability, simply will not work as well. Studies do indicate that when devout personal religious commitment is combined with active participation and involvement in the religious community, health effects are magnified." Much of that evidence is outlined in Appendix 2.

Confirmation of the health benefits of religion would raise many intriguing issues. First, would doctors prescribe a faith? And if they did, would it really count as faith (at least in the eyes of religious leaders) if someone chose a given religion simply because they wanted to live longer rather than because they believed in a certain doctrine? And would this discovery mark a return to that elusive Christmas of yesteryear, when faith, not consumerism, held center stage?

Science versus Christmas

There is one last issue to be dealt with. Does the scientific worldview somehow undermine the religious beliefs that are the basis of Christmas for so many people? Science has been viewed suspiciously as a force that turned people away from God ever since 1916, when an oft-cited survey by James Leuba of Bryn Mawr University found that 60 percent of American scientists did not believe in God. The finding caused a scandal at the time, prompting warnings from politicians about the evils of modernism and accusations that scientists were leading college students away from religion. Leuba himself predicted that disbelief among scientists would only increase in the future.

But research conducted recently, repeating the 1916 survey word for word, has proved Leuba wrong. The proportion of scientists who believe in God has remained almost unchanged in the past eighty years, despite the enormous leaps of discovery made during this century. Edward Larson, from the University of Georgia, and colleague Larry Witham, from Burtonsville, Maryland, questioned six hundred scientists listed in the 1995 edition of *American Men and Women of Science* and achieved the same results as Leuba: about 40 percent of scientists believe in God. The future of Christmas and Hanukkah in our increasingly technological age seems assured.

11

SANTA'S SCIENCE

I would stay awake all the moonlit, snowlit night to hear the roof-alighting reindeer and see the hollied boot descend through soot. But soon the sand of the snow drifted into my eyes, and, though I stared towards the fireplace and around the flickering room where the black sack-like stocking hung, I was asleep before the chimney trembled and the room was red and white with Christmas. But in the morning, though no snow melted on the bedroom floor, the stocking bulged and brimmed; press it, it squeaked like a mouse-in-a-box; it smelt of tangerine; a furry arm lolled over, like the arm of a kangaroo out of its mother's belly.

DYLAN THOMAS,
"CONVERSATION ABOUT CHRISTMAS"

IT ALL LOOKS SO EFFORTLESS. APART FROM THE occasional slipup with drunken reindeer, narrow chimneys, and blizzards, Santa manages to deliver millions of gifts on Christmas Eve, maintaining his smile and composure all the while. His support team: a few reindeer and a handful of diligent elves.

I beg to differ. Only an innocent child would swallow this propaganda, a fantasy peddled by generations of Christmas cards to divert attention away from what is, undoubtedly, the most spectacular research and development outfit this planet has ever seen. I have good reason to believe that Santa has drawn on the benefits of centuries of inventions and insights generated by a scientific effort that would make the likes of Albert Einstein weep with admiration.

Somewhere in the North Pole, or perhaps buried in a vast complex under Gemiler, original home of Saint Nicholas, there must be an army of scientists experimenting with the latest in high-temperature materials, genetic computing technologies, and warped space-time geometries, all united by a single purpose: making millions of children happy each and every Christmas.

Put yourself in Santa's boots: How does he know where children live and what gifts they want? How can he fly in any weather, circle the globe overnight, carry millions of pounds of cargo, and make silent rooftop landings with pinpoint accuracy?

Some years ago *Spy* magazine examined these issues in an article that has since proliferated across the Internet. The piece concluded that Santa required 214,200 reindeer and, with his huge mass of presents, encountered "enormous air resistance, heating the reindeer up in the same fashion as a spacecraft re-entering the earth's atmosphere." In short, the article concluded, the reindeer "will burst into flame almost instantaneously, creating deafening sonic booms in their wake. The entire reindeer team will be vaporized within 4.26 thousandths of a second. Santa, meanwhile, will be subjected to forces 17,500.06 times greater than gravity. . . . If Santa ever DID deliver presents on Christmas Eve, he's dead now."

The point is that Santa is *not* dead. He delivers presents every Christmas Eve, as reliably as Rudolph's nose is red.

And he overcomes the kinds of problems outlined above with the aid of out-of-this-world technology. Let's examine just how he does it.

Santa's Challenge

Santa has a huge market. There are 2,106 million children under age eighteen in the world, according to the United Nations Children's Fund (UNICEF). Given the pagan origins of the festival and the holiday's emphasis on charity, I will assume that Santa delivers presents to each and every child, and not just to Christian children or to the 191 million who live in industrialized countries. It *is* Christmas, after all.

If we assume that there are 2.5 children per household, Santa has to make 842 million stops on Christmas Eve. Now let's say these homes are spread equally across the landmasses of the planet. Earth's surface area is, given a radius of 3,986 miles (6,378 kilometers), 196.6 million square miles. Only 29 percent of the surface of the planet is land, so this reduces the populated area to 57.9 million square miles. Each household covers an area of 0.069 square miles. Let's assume that each home occupies a square plot, so the distance between households is the square root of the area, which is 0.26 miles.

Every Christmas Eve Santa has to travel a distance equivalent to the number of chimneys — 842 million — multiplied by this average spacing between households, which works out to be 221 million miles. This sounds daunting, particularly given that he must cover the entire distance in one night.

Fortunately, Santa has more than twenty-four hours to deliver the presents. Consider the first point on the planet to go through the international date line at midnight on December 24. From that moment on, Santa can pop down chimneys. If he stayed right there, he would have 24 hours to deliver presents to everyone along the date line. But he can do better by traveling backward against the direction of Earth's rotation. That way he can deliver presents for almost another 24 hours to everywhere else on Earth — making 48 hours in all, which is 2,880 minutes or 172,800 seconds.

From this one can calculate that Santa has a little over 2/10,000 second to get between the 842 million households. To cover the total distance of 221 million miles in this time means that his sleigh is moving at an average of 1,279 miles per second. Ignoring quibbles about air temperature and humidity, the speed of sound is something like 750 miles per hour, or 0.2 miles per second, so Santa is achieving speeds of around 6,395 times the speed of sound, or Mach 6,395.

When any object exceeds the speed of sound, there will be at least one sonic boom. This is a shock wave sent out when the sleigh catches up with pressure waves it generates while moving, explains Nigel Weatherill of the University of Wales, Swansea, who helped the Thrust Supersonic Car break the sound barrier in 1997. But Santa does not generate any sonic booms on Christmas Eve. In his book *Unweaving the Rainbow,* Richard Dawkins says he has used this fact to disprove the existence of Santa to a six-year-old child. To a biologist this may indeed seem persuasive. But to aerodynamics engineers, it suggests that Santa has found a way to suppress sonic booms. For example, says Weatherill, perhaps Santa cancels the peaks and troughs in the shock wave with troughs and peaks of "antisound" generated by a specialized speaker on his sleigh.

The speed of light is absolute and cannot be exceeded (however, see below), so we better check to make sure Santa is not breaking this cosmic speed limit. The usual figure quoted for the speed of light is 300 million meters per second, which works out to be 186,000 miles per second. Santa is comfortably within the limit, traveling at around 1/145 the speed of light — too slow to worry about the implications of Einstein's theory of relativity.

The preceding discussion assumes that Santa throws the presents down each chimney while whizzing overhead. In fact, he stops at each house, so he has to achieve double the speed calculated above. From a standing start, he has to travel the distance between houses in 2/10,000 second. That means going from 0 to 2,558 miles per second in 2/10,000 second, an acceleration of 12.79 million miles per second per second, or 20.5 billion meters per second per second.

The acceleration due to gravity is a mere 9.8 meters per second per second, so the acceleration of Santa's sleigh is equivalent to about 2 billion times that caused by the gravitational tug of Earth. Given that Santa is somewhat overweight, say around 200 kilograms, the force he feels is his mass times his acceleration: around 4,000 billion newtons. Even fighter pilots can't cope with accelerations more than a few times that of gravity, and they have to use special breathing techniques and G (gravity) suits to keep the blood in their heads. As the physics professor Lawrence Krauss puts it, the acceleration Santa has to cope with would normally reduce a person to "chunky salsa."

Krauss has considered similar problems in his book *The Physics of Star Trek*. The starship *Enterprise* gets by with devices called inertial dampers to cushion the forces that Captain Kirk feels in the seat of his pants. Santa has to resort to similar tactics, creating an artificial world within his sleigh in which the reaction force that responds to the accelerating force is canceled, perhaps by some kind of elf-made gravitational field.

There is one other problem Santa has to contend with: his cargo. Assuming that each of the 2,106 million children gets nothing more than a medium-size construction set (2 pounds, or 0.9 kilograms), he has a payload of 4,212 million pounds (about 2 million tons), or 1,895 million kilograms, of toys. Then there is the supply of fuel required to achieve the high speeds he must maintain. Any way you look at it, Santa has some serious hurdles to overcome.

Einstein and Santa

So much for the challenge facing Santa. What actually goes on when he delivers all those presents? As an academic exercise in fun (and a gift to beleaguered parents, let alone journalists short of hard copy around the holiday season), Larry Silverberg at North Carolina State University decided to deliver some scientific answers to all those troublesome questions about Santa. He is no fool, after all, but a professor of mechanical and aerospace engineering and a member of NASA's Mars Mission Research Center at the university.

Assisting Silverberg were fellow North Carolinians Robert Stanley, a doctoral student in mechanical engineering; J. P. Thrower, a doctoral student in electrical engineering; Dan Deaton, a senior in mechanical engineering; Charles Grant, a senior in mechanical engineering; and Jeffry Stock-Windsor, a doctoral student in mechanical engineering.

Let's begin with the sleigh. Silverberg describes it as one of the engineering wonders of the world and thinks he knows how it works — enough so to become a true believer.

"Santa clearly is ahead of the curve when it comes to applying advanced scientific theories to his sleigh's design. Children shouldn't believe others who say he isn't real because there's no way he could deliver toys all over the world in one night. There *is* a way, and it's based on plausible science."

Silverberg suggests that Santa exploits features of Einstein's theory of relativity. The first part, special relativity, starts from the premise that the speed of light and the laws of physics are the same (invariant) for observers moving at constant speeds relative to one another. The bottom line of special relativity is that common sense falls apart when we travel at high speeds, particularly when we approach the ultimate speed, that of light. The theory says that the speed of light is constant, no matter the position or speed of an observer.

That is why relativity seems so odd. A snowball thrown by Santa as his sleigh whizzes past would seem to him to move slowly compared with the Earth-bound perception of the snowball's speed. From our vantage point on the ground, the snowball's velocity includes that of the sleigh. But relativity says that light breaks this commonsense rule. The speed of light from a flashlight turned on by Santa looks the same to him, from his vantage point in the moving sleigh, as to someone standing below in the snow.

In our everyday experience, it is time and space that are absolute; inches and seconds are the same, wherever you are on the planet or whatever you are doing. But the only way observers in relative motion can come up with the same speed of light — say, for a laser beam from a Christmas light display to pass from one planet to another in a given time period — is if each of their "seconds" or each of their "meters" is different. The higher Santa's speed, the more time apparently dilates and space contracts, Silverberg says.

Special relativity gives Santa ample opportunity, within his frame of reference, to deliver those presents in what is an eye blink by our frame of reference, Silverberg argues. "In his reference frame, time moves much quicker than in our own — that is when he is inside his 'relativity cloud.' In thc relativity cloud, because time moves much slower for us [relatively], he sees us basically frozen. He doesn't even need to hurry. He has all the time in the world."

The next difficulty to overcome is to show how Santa achieves such high speeds. Even if a rocket sleigh could get him to that small fraction of light speed, the fuel requirements would be prohibitive. This is where we need the second part of relativity theory, or general relativity, which broadens Einstein's original theory to accommodate the idea that the laws of physics should be the same for all observers, regardless of how they are moving relative to each other.

General relativity was conceived after Einstein realized that if a person fell off a roof, he would not feel his own weight — until he hit the ground. This theory replaced Newton's way of describing gravity — that is, not as a force, but as the curvature of space-time, a four-dimensional mix of space and time. For example, Earth orbits around the sun because space-time has been warped into a shape like the inside of a bell, with the sun sitting at its center.

Silverberg points out that although our theoretical understanding of space-time was laid down during the first decades of this century, our fat friend is an old hand at such science. "Santa and his community at the North Pole have known about this much longer. In fact, they have learned, as a practical matter, how to manipulate time, space, and light — to the extent that they can control these phenomena. With their hundreds of years of experience with understanding relativity, they have turned it into a working theory; they've made relativity clouds that fit Santa, his sleigh, and all of the reindeer."

According to relativity theory, matter cannot move through space faster than the speed of light. But there is no limit on the speed at which space itself can move. The sleigh can sit at rest in a small bubble of space that flows at superluminal velocities through normal space. Another way to put it is that objects can travel locally at slow velocities yet be traveling faster than the speed of light because the very fabric of space-time is stretching.

A proof of the principle of this idea was put forward in 1994 by Miguel Alcubierre while at the University of Wales in Cardiff. He demonstrated that this apparent loophole is entirely consistent with modern physics and that Santa's sleigh could indeed travel faster than light.

Alcubierre found that if space-time can be warped locally so that it expands behind Santa's sleigh and contracts in front, the craft will be propelled along with the space it is in, riding the crest of the wave. By his scheme, the space-time around the sleigh is being warped so that it can be moved between chimneys without experiencing much, if any, local acceleration. Instead, the space-time between Santa and the last chimney is expanding, bringing the next one closer.

The sleigh will never travel locally faster than the speed of light, because the light, too, will be carried along with the expanding wave of space. However, the accompanying wave will allow Santa to travel huge distances in almost no time at all, let alone the short distance between chimneys.

Santa can go even further and cut and paste different bits of the universe together, using space-time shortcuts called wormholes, an idea explored by the mathematician Ian Stewart at Warwick University in England. If you imagine space-time as being curved, like a sheet, a wormhole offers a more direct route than following all the curves and contours of the sheet, going outside the normal universe. Santa enters one portal, passes through the wormhole, and emerges from another. He can carry one end of the wormhole on his vehicle and arrange for the other to materialize inside each dwelling that he visits. No more sooty chimneys or getting stuck inside central heating systems, Stewart says.

The wormhole idea also can allow time travel, providing unlimited opportunities to deliver an annual cargo of presents. This possibility was raised in a 1988 paper in *Physical Review Letters* by Michael Morris, Kip Thorne, and Ulvi Yurtsever and rests on the so-called twin paradox. Imagine two identical reindeer, Donner and Blitzen, Stewart says. Donner remains on Earth, and Blitzen heads off into space at nearly light speed, returning forty years later as measured by an Earth-bound observer. Donner has aged forty years, but because of time dilation, Blitzen has aged only a few.

Morris, Thorne, and Yurtsever realized that by combining a wormhole with the twin paradox, they could get a time machine. The idea is to leave one end of the wormhole fixed and to zigzag the other end to and fro at just below the speed of light. Seen from inside the wormhole, both ends age at the same rate. But from outside, the moving end ages more slowly because of its speed. From the perspective of an elf hanging about the fixed wormhole portal, a daylong journey of the moving end could appear to take ten days.

But if the curious elf were to peer through the wormhole, he would see things as they were nine days earlier. This means that the passage of time is different if you wriggle from one end of the wormhole to the other. Indeed, if you travel through normal space to the moving end and then dive through the wormhole, you end up in your past.

There are problems to overcome here. For example, time travel risks creating paradoxes that defy commonsense ideas about cause and effect. The classic one is the grandfather paradox, where Santa goes back in time and runs over his three-year-old grandfather, with fatal consequences for himself.

One way to sidestep such paradoxes may come from the many-worlds interpretation of quantum mechanics, the lore that rules the subatomic world. This controversial interpretation of quantum mechanics is, in the words of one wag, cheap on assumptions and expensive on universes. If you believe it (and many don't), each journey through time carries us into a new version of the universe, one that coexists with the original universe but is separated from it along some totally new kind of dimension. There could be trillions of Santas out there delivering presents in parallel universes.

However, there is a deeper problem with the preceding discussion. In all Alcubierre's talk of weird and warped space-time that may allow Santa to travel at light speed and beyond, I cunningly overlooked how Santa manages to mangle the fabric of the universe in the first place.

According to Alcubierre, warping space-time would require the presence of exotic matter that, unlike the stuff we are familiar with, repels other matter by exerting an antigravity effect. That is at least according to the theory of relativity. Quantum mechanics does allow for the existence of "negative energy" matter at the scale of atoms and molecules. But the big question is whether matter can have this property on the scale of reindeers and sleighs.

A recent calculation investigated this very issue. Mitchell Pfenning and Larry Ford at Tufts University found that this energy would have to be packed into a doughnut-shaped region wrapped around the sleigh. However, they calculated that the total amount of energy needed to sustain the warp would be huge — around ten billion times the energy locked up in all the visible mass in the universe. That would seem to rule out warp-drive sleighs.

But it may not be over for Santa, according to Alcubierre. Pfenning and Ford's work is based on approximations that, strictly speaking, work only in space that is not already warped. We lack the theory — a fusion of quantum mechanics and relativity — to carry out this calculation accurately at present. Has Santa learned how to exploit this next generation of theory, so-called quantum gravity? Given what happens each Christmas Eve, this is entirely possible.

The Traveling Santa Problem

Many other questions remain for us to tackle. How does Santa know where children live and what gifts they want? Although old-fashioned letters to Santa still work, Silverberg and his students speculate that he relies on a strategically placed multigrid antenna system that picks up electromagnetic signals from children's brains.

There exists, after all, a technique called magnetoencephalography, which uses a Squid (a superconducting quantum interference device) to detect minute magnetic fields generated by the crackle of brain activity. Sophisticated signal-processing methods are then used by Santa to filter the data and ascertain who the children are, where they live, and whether they've been bad or good. This data is transferred to an onboard sleigh guidance system, which uses a computer to plan the most efficient route of delivery.

Here, of course, Santa runs into a classic quandary: the traveling salesman problem. This is the dilemma faced by the salesman who has to visit a number of cities, each only once and in a way so as to minimize the total distance traveled. (He has a penny-pinching boss and has to keep fuel costs as low as possible.) In the festive season many scientific minds turn to solve that perennial chestnut, the "traveling Santa problem."

For a handful of cities and roads, it may be easy to determine the salesman's shortest route, because not that many options exist. If the number of cities is 5, a computer could easily calculate the 120 possibilities. With 10 cities there would be 3,628,800 possibilities.

However, even with the number-crunching power of the fastest available machine, the time required to solve this type of problem rapidly spirals out of control. For just 25 cities the number of possibilities is so immense that a computer evaluating a million possibilities per second would take 490 billion years — that is, about 40 times the age of the universe — to search through them all. For the 842 million households Santa has to visit it would take 10 to the power of 7.15 billion years. Now that would certainly present Santa with a headache if he were trying to be efficient.

Larry Silverberg points out that present-day computers could calculate a route that is 99.99 percent "near" the best possible solution — which would be sufficient for Santa's purposes. The traveling Santa problem is tackled by first being broken down. Cities are grouped into countries and weighted according to the number of children in the country.

Then each city is given a weighting number that corresponds to the number of children in that city, and so on. "We start by optimizing by country and then by city and finally by home," Silverberg says. "We can perform more subdivisions when needed, but in the end the number of calculations can be performed using present-day, albeit fast, computers. When you call a friend, telephone companies use switching algorithms that are related to these kinds of suboptimal solutions."

Given Santa's extraordinary research and development operation, he probably has a DNA computer, an idea pioneered by Leonard Adleman of the University of Southern California. In a milestone paper Adleman tackled something called the "directed Hamiltonian path problem," which also involves finding a special path through a network of points. We can think of this in the same terms as the problem facing Santa.

The computer code, Adleman proposed, consisted of a mixture of trillions of pieces of single-stranded DNA, each piece representing a city or a route. Among the vast number of combinations that resulted from the binding together of complementary DNA strands, it was overwhelmingly probable that one combination corresponded to the solution sought. Standard molecular biological techniques were then used to fish that molecule out: the "solution" could be distinguished by its length and the details of its composition.

Whereas current supercomputers are able to perform a million million million operations per second, molecular computers could conceivably run billions of times faster. That is fast enough, when combined with mathematical cunning, for Santa to plan the most efficient route.

But, alas, his challenges continue. How can one sleigh

carry two million tons of presents? It doesn't have to, explains Silverberg. To tackle this problem, one dear to the heart of every child, he invokes nanotechnology, an idea first put forward by the late Nobel laureate Richard Feynman.

Santa can now use a hierarchical distributed mobile manufacturing system to make the gifts on-site in each child's home. Silicon chip–based machines, so tiny they can fit on the head of a pin, are loaded with a code containing the child's toy list. The machines literally grow the toys, atom by atom, from soot and snow and other chemicals Santa collects along his route. Large toys require thousands of nanomachines working in concert and can drain Santa's technological resources, Silverberg warns, which is why children shouldn't expect more than one big gift each year.

DNA computers, nanotechnology, and warped spacetime seem plausible enough, but how do reindeer fly? The ability is in their genes, argues Silverberg. After centuries of selective breeding and (more recently) bioengineering, the volumetric displacement of their lungs is of such a proportion that when filled with an appropriate mixture of helium, oxygen, and nitrogen, they become buoyant. Pulling the sleigh becomes as easy for them as pulling a raft in water.

The reindeer could also be powered by jetpacks, Silverberg adds. He cited a soccer game that was "kicked off" with a mascot strapped to jetpacks that touched down in the middle of a stadium. "In fact, the reindeer don't even have to be lightweight — although it helps. Our team at NC State speculates that Santa uses reindeer simply because they are his favorite Arctic animal. Also, they have good balance, which is needed when landing on rooftops."

Despite this natural ability, the reindeer may require little stabilizers to keep Santa, the sleigh, and all those presents on an even keel. In studies of fruit flies, scientists have discovered body plan genes, called homeotic genes, that may enable Santa to do just that. Amazingly, a mutation in one of these genes can transform one body part to another, such as an antenna to a leg, a thorax to a wing, or a nose to a leg.

Reindeer could be customized for flight by manipulating these homeotic genes, speculates Matthew Freeman, a geneticist at the Laboratory of Molecular Biology in Cambridge, England. "Work by Cliff Tabin of Harvard Medical School and others has shown that genes that specify limbs in flies, such as wings and legs, are the same as in mammals and even work in much the same way. That means that, with Santa's genetic wizardry, Rudolph could grow stabilizers or even wings," Freeman says.

Stewart adds that we may have underestimated the importance of one of Rudolph's features: his headgear. He points out that the aerodynamics of hypersonic flight is highly counterintuitive anyway, though perhaps no more so than the good old heavier-than-air flying machines we now take for granted. "Reindeer have a curious arrangement of gadgetry on top of their heads which we call antlers and naively assume exist for the males to do battle to win females," he says. "This is absolute nonsense. The antlers are actually fractal vortex-shedding devices. We are talking not aerodynamics here, but antlaerodynamics."

At supersonic speeds, the Concorde generates lift by shedding a big helical vortex from the tip of each wing. The reindeer, at their much higher speeds, play the same trick with the prongs of their antlers, Stewart explains. "These spin off a whole system of vortices, carefully tailored to generate the right amount of lift for such high speeds. [You don't need much!] So the reindeer hang from their antlers as they fly — which is why the antlers are on top and at the front."

No doubt Santa used supercomputers to design the optimal antlaerodynamic configuration and genetic engineering to produce it in the actual reindeer. Just as penguins look comical on land but come into their own underwater, so the apparently Earth-bound reindeer reveal the true beauty of their design only when traveling at speeds in excess of Mach 6,000. "It may be that the reindeer shed their antlers each year in order to upgrade their aerodynamics to the latest technology," Stewart says, though he adds, "I'm not certain of this."

Of course, if the reindeer traveled *too* fast, they could burn up, along with the sleigh. To provide adequate thermal shielding — of both reindeer and warp-drive sleigh — the team at North Carolina proposed a Kevlar-like composite fiber encapsulated in an epoxy resin matrix. This would make the shield very strong, lightweight, durable, and cold resistant. Other possibilities include the silica materials used on the space shuttle or materials called aerogels.

Anecdotal evidence also backs the idea that Santa uses heat shields, Larry Silverberg notes. "When going through the ionosphere, a sleigh made of this material would glow like a speck of bright red, a sight children and other reliable sources have long reported seeing in the sky on Christmas Eve." This is just one of a range of achievements that ensure children the world over are not disappointed on Christmas morning.

And so we have it. Santa's science has, at last, been laid bare.

12

CHRISTMAS 2020

"Ghost of the Future!" he exclaimed, "I fear you more than any Spectre I have seen. But as I know your purpose is to do me good, and as I hope to live to be another man from what I was, I am prepared to bear your company, and do it with a thankful heart. Will you not speak to me?"

CHARLES DICKENS, A CHRISTMAS CAROL

I WAS AWOKEN BY A SHAFT OF LIGHT FROM MY bedside table lamp. Without my morning dose of daylight-intensity photons, I would have been even more depressed about Christmas Day than usual. The flat-screen TV hanging on the wall crackled into action to give me weather diagnostics, digital Yuletide greetings courtesy of a cola manufacturer, and a glimpse of the world outside my 195th-floor apartment.

The snow lay deep and crisp and even, just as the atmospheric engineers said it would. "Silent Night" was blaring from a passing Santa-shaped dirigible as its digital billboards exhorted the masses to finish their last-minute teleshopping. And an e-mail told me the local transgenic farm had just delivered a three-pound, high-fiber, cloned turkey breast fillet, engineered with a mixture of tenderizing enzymes and Maillard reagents to improve taste and color.

A hum sounded nearby. My 3-D fax was delivering another gift sent over the web. Once I had visited my sanichamber and gargled with my cinnamon-flavored plaque-digesting enzymes, I decided to take a peek. Through the quartz glass portal of the 3-D fax I could see a number of sparkles of red, blue, and green light as laser beams crossed in the vapor, building up polymer deposits into another present. I yawned. It looked as if it was going to be one of those designer Santa effigies you could personalize with the sender's or recipient's face.

Fortunately, the neural net software in my message center had already figured out whom the gift was from, assessed its value down to the last euro, and used a game theoretic software package to select the appropriate reciprocal gift to send back via the 3-D fax. All that, and a personal message of thanks, lovingly crafted by fuzzy logic software, then signed with the appropriate seasonal wishes: Happy Kwanzaa/Christmas/Hanukkah/Holidays. Thank goodness we don't have to think about what to give people anymore.

I put on my virtual reality headset to check out the latest store offers before lunch. "Click-on, cruise-thru" a mall web site and pile your cyber-shopping trolleys high with the latest technomagic! Experts used to predict that Christmas shopping at virtual stores would be a lonely affair. But now that everyone could cruise around as an avatar, a digital alter ego, it was fun to see the last-minute frenzy without all the physical hassle, shopping cart rage, and fights over the last remaining turkey or mince pie.

Look at that, I thought. A final discount on the latest cloned firs was broadcast. They were Gucci trees, teardrop-shaped with a designer logo *and* a guarantee that not a single needle would fall on your self-cleaning floor. Two eye blinks and a nod of the head later, I started to examine the goods. One began to rotate. I breathed in. The odor chip in the headset sensed this sharp intake of conditioned air and did a passable imitation of the tree's smell, down to the last whiff of alpha pinene. "I'll take it."

After the usual Bloomingdale's shopbot formalities, quantum encrypted credit checks, and extras (no, I did not want the tree's jellyfish gene activated — glowing trees are *so* vulgar), my customized tree dropped into my cart, complete with black fairy. Twenty thousand euros, a three-hour delivery from tissue plant to door, plus a trade-in for my old biodegradable plastic tree. It was a bargain.

Off with the VR helmet and into the kitchen. The turkey had already been screened for harmful bacteria with a gene chip and would take only a few minutes to deal with. The smart cooker had selected the appropriate vegetables, seasonings, and so on. The Shiraz was on the table, ready to drink. This time I had made sure the wine would not interfere with any mood-altering drugs by selecting one of the low-alcohol engineered reds from Australia.

Armed with my body mass and fat index — updated by my brief visit to the sanichamber — the small pack of tabletop electronics in the Dietmate had already worked out what would be healthy to eat and started to mutter something about cutting the calories by boiling the potatoes. I decided to overrule it today and issued the command, "Roast them in drippings!" The Dietmate began to plan low-fat retribution for the New Year.

Later came the highlight of the celebrations: a Highfield family virtual reunion. This year we had consulted an Imagineer at one of Disney's virtual worlds, among the best-known providers of on-line gatherings. For our get-together we had asked for "something like Santa's grotto" with the full effects — smellyvision, virtual celebs, the works. There were the usual jokey avatars. I went for Thomas Nast–style Santa garb and a seasonal cinnamon smell, and I selected a winged Rudolph to get me around. The motion feedback chair made me feel a little seasick when my charger started to fly.

There were other surprises. Photos of some long-dead, much-loved relatives, scraps from their voice mail, and a few family anecdotes cobbled together by the Imagineer and processed by computer. Though long gone, their digitally re-mastered bodies lived on to send a brief Christmas message and even answer a question or two at the family reunion. Using black-and-white images and a crackly sound track, the effect was eerie. I bet a few VR headsets misted up after that.

Earlier that week the family elders had spent some time telling the Imagineer who got along with whom and conveying details of the worst family feuds. That way the high-level software could distract us with a virtual celebrity, even march us off with a virtual elf if there was any risk of a clash between avatars. On-line feuds can be much worse than the real thing. Once you have rowed in cyberspace, you can't easily forgive and forget. Victims replay and brood, replay and brood until the next Christmas. Just in case there were any clashes, we took the usual pharmacological fixes — semiochemicals, dopamine boosters, serotonin blockers — to tweak the neurotransmitters for the full festive mood.

And of course, we selected *A Christmas Carol* for those who wanted to attend but sought a refuge from some of the relatives. There are a lot of options available in these digitally customized films — too many. This had prompted the usual flurry of e-mail, voice mail, and picmail during the days before. We all agreed on a period setting, then squabbled about who should play whom. In the end we adopted a compromise, with Alec Guinness as Scrooge and Stan Laurel playing a particularly pathetic Cratchit. The supercomputer did the rest. For the sound track we had data-mined one of the entertainment archives, using a neural net to cross-reference our personal tastes with appropriate music. Then all we had to do was settle down to watch in our VR headsets.

A synthetic orchestra and choir began to play a carol, and a ghostly image of Marley appeared. As I sat under my cloned tree, I swallowed a smart pill to try to help unlock some of my childhood memories. The snow, seeded with silver iodide, continued to fall outside. I felt a warm glow inside. You can't beat a traditional Christmas.

APPENDIX 1

The Formula for Christmas Day

For all those who like to plan their Christmas in advance, here is a little formula that can reveal the day of the week that Christmas Day falls on in any year (including leap years) after 1600.

1. Write down the year you're interested in, say 1998. Split this number into its century number, C (19, in this case), and its year number, Y (98 here).
2. Now divide C by 4 and keep just the whole-number part of the result, which we will call K. In this case, $K = 19 \div 4 = 4.75$, which is rounded off to 4.
3. Do the same for Y, giving a new figure, G. In this example, $G = 98 \div 4 = 24.5$, rounded off to 24.
4. Now work out the value of D, using the formula

   $$D = 50 + Y + K + G - (2 \times C)$$

   In our case,

   $$D = 50 + 98 + 4 + 24 - 38 = 138$$

5. Finally, to work out the day of the week that Christmas falls on, divide D by 7 and write down the remainder, R. Use the following table to give the day of the week:

R = 0 Sunday
R = 1 Monday
. . . . .
R = 6 Saturday

In our case, $D = 138$, so D divided by 7 is 19 with a remainder of 5. This means Christmas Day 1998 will fall on a Friday. The same formula shows, for example, that Christmas Day in the year 2164 will fall on a Tuesday.

APPENDIX 2

Is Faith Good for You?

There is growing evidence that those who respect religious traditions, and presumably those who take Christmas seriously, can expect a healthier life.

A wide-ranging survey of scientific evidence of the "faith factor" in disease has been conducted by Dale Matthews of the Georgetown University School of Medicine. He reviewed 212 studies and found that religion had a positive effect in three-quarters of them, notably those investigating substance abuse, alcoholism, mental illness, quality of life, illness, and survival. "Whether one looks at cancer, hypertension, heart disease, or physical functioning, most of the studies demonstrate a positive effect," he said. Seven out of ten studies on survival showed that religious people live longer.

Among the investigations Matthews cited are the following:

- A survey of 91,909 individuals in Washington County, Maryland, found 50 percent fewer deaths from coronary artery disease, 56 percent fewer deaths from emphysema, 74 percent fewer deaths from cirrhosis, and 53 percent fewer suicides in religious people.

- A landmark study of 4,725 people in Alameda County, California, showed that church members had lower mortality rates than others, independent of socioeconomic status and smoking, drinking, physical inactivity, and obesity.
- A study of 522 Seventh-Day Adventist deaths in the Netherlands revealed they had an additional life expectancy of nine years for men and four years for women when compared with the general population. Although the group is vegetarian, Matthews added that "the reason that they are vegetarian is their religion."
- Mormons enjoy unusually good health, with cancer and heart disease rates less than one-half those of the general population. Highly religious Mormons in Utah experience one-half the rate of cancer, as do less adherent members of the faith, even though they constitute a highly homogeneous social group.
- Among 1,344 outpatients in Glasgow, those who participated in a religious activity at least monthly were less likely to report physical, mental, and social symptoms.
- A study of 1,400 women in Michigan found that church attendance was correlated with a longer life, while "TV watching was associated with death," said Matthews. "This presumably is a variable for lack of physical activity, but you never can tell."
- And almost all patients who undergo heart surgery — including atheists — pray. A study of bypass surgery patients showed that the six-month mortality rate was 9 percent overall, 5 percent in churchgoers, 12 percent in those who did not go to church, and zero in the "deeply religious."

Since Matthews's roundup, there has been other support for the faith factor, notably two reports, coauthored by Ellen Idler, at Rutgers's Institute for Health, Health Care Policy, and Aging Research, and Stanislav Kasl, of Yale University School of Medicine. The papers, published in the *Journal of Gerontology*, reveal the findings of a twelve-year study that sampled 2,812 people age sixty-five and over from Protestant, Catholic, Jewish, and other religious backgrounds in New Haven, Connecticut.

The first study found a lower frequency of unhealthy behaviors; better support systems and social ties; and improved emotional well-being. The second report explored how religious involvement can influence changes in physical health over a twelve-year period, revealing that attendance at religious services was a good predictor of functional ability in later life.

"Over the long term people who had better health levels in 1982 and continued attending religious services were able to maintain higher levels of functioning and psychological health through 1988," Idler remarks. "Even after we took out the other variables such as friendship, leisure activities, and social support, there was still evidence that attendance at religious services had a positive impact.

"There were so many reasons for thinking that we should expect better health among people who are religiously involved, but until now it wasn't anything we were able to quantify," she says. "We also found that it wasn't a person's individual feelings of religiousness that made the difference; it was acting as part of the larger worship group that fostered positive health."

Patients are to some extent driving this interest in the link between faith and health. A 1994 study found that 73 percent of hospital inpatients prayed on a daily basis, 77 percent said that doctors should consider their spiritual needs, and 48 percent wanted their doctors to pray with them.

Doctors, however, are cool to the idea of the faith factor, according to Kenneth Pargament of Bowling Green State University. A study of 2,348 psychiatric surveys conducted over five years found that few collected basic data on religious belief: only 2.5 percent used a religious measure and only 0.1 percent had religion as a central variable, with only one employing a validated measure of religious commitment. Pargament goes so far as to argue that health research that ignores religion is incomplete. "Faced with the insurmountable, the language of the sacred — hope, surrender, forgiveness, serenity, divine purpose — becomes more relevant," he says. "Ultimate control is still possible through the sacred when life seems out of control."

One reason that doctors are not swayed by a simple picture of the "faith factor" is that certain forms of religious expression are clearly counterproductive for better health. One has only to think of the tragedies surrounding the Jonestown, Waco, and solar temple cults. In addition, some groups reject medical care, such as the Faith Assembly in Indiana, which has three times the state average perinatal mortality rate and one hundred times the maternal mortality rate because of lack of obstetric care.

"When you look more closely, you find there are certain types of religious expression that seem to be helpful and certain types that seem to be harmful," Pargament says. In several studies involving hundreds of subjects, he has found that people who embrace what could be called "the sinners in the hands of an angry God" model do indeed have poorer mental health outcomes. People who feel hostility toward God, believe they're being punished for their sins, or perceive a lack of emotional support from their church or synagogue typically suffer more distress, anxiety, and depression.

By contrast, people who embrace the "loving God" model see God as a partner who works with them to resolve problems. They view difficult situations as opportunities for spiritual growth. And they believe that their religious leaders and fellow congregation members give them the support they need. The result? They enjoy more positive mental health outcomes, Pargament says.

Support for this view has come from the work of Lee Kirkpatrick, at the College of William and Mary, who discovered that people who viewed God as a warm, caring, and dependable friend were much more likely to have positive outcomes than people who viewed God as a cold, vengeful, and unresponsive deity or who weren't sure whether or not to trust God. "People who classified their attachment to God as secure scored much lower on loneliness, depression, and anxiety and much higher on general life satisfaction," Kirkpatrick explains.

GLOSSARY

ABSOLUTE ZERO: No Christmas can get colder than this. The lowest conceivable temperature is –273.16°C. (For comparison, a deep freezer is around –15°C.) A fundamental law of physics, known as the third law of thermodynamics, says that nothing can actually be cooled down to absolute zero. Scientists can now come to within a few billionths of a degree of this baseline. As something cools, the average energy and movement of its molecules drops. You might think they would come close to stopping altogether near absolute zero, but quantum theory says that even at such extremely low temperatures, molecules cannot help but jiggle about a bit.

AMINO ACIDS: The molecular building blocks of proteins.

ATOMS: Atoms were regarded by the Greeks as indivisible units of matter. Now they are seen as the smallest units that bear the chemical characteristics of an element, whether hydrogen or uranium. More than a billion atoms would fit on the period at the end of this sentence. They are mostly empty space. The nucleus, where most mass resides, is one hundred thousand times smaller than the overall atom. Each atom consists of a positively charged nucleus orbited by a mist of negative charge.

Cell: A discrete, membrane-bound portion of living matter; the smallest unit capable of an independent existence.

Chaos: A term often used to describe preparations for Christmas. In science it is used to describe apparently random behavior. The branch of mathematics that deals with such behavior is called chaos theory. The essence of chaos is expressed in the butterfly effect: a butterfly flapping its wings near Santa's grotto can cause a subsequent hurricane over Texas. Earth's atmosphere is so sensitive that if there is the even slightest uncertainty in the current weather conditions, the weather within a few weeks' time, such as a white Christmas Day, is unpredictable.

Christmas: "A day set apart and consecrated to gluttony, drunkenness, maudlin sentiment, gift taking, public dullness and domestic behavior" (Ambrose Bierce).

Chromosome: A long strand of DNA containing thousands of genes in a package. There are twenty-three pairs in all human cells, except eggs and sperm.

Cross-linking: The process that makes a turkey tough. The formation of side bonds between different chains in a polymer, such as the proteins in meat, increasing its rigidity.

Determinism: The theory that a given set of circumstances inevitably produces the same consequences, rather like exuberant office parties and hangovers.

DNA (deoxyribonucleic acid): The vehicle of inheritance, from reindeer to wise men. DNA is a complex giant nucleic acid molecule carrying the genetic blueprint for the design and assembly of proteins, the basic building blocks of life. DNA, the double helix, is a chainlike molecule made up of a series of bases that come in four flavors (adenine, guanine, cytosine, and thymine, or A, G, C, and T). The order of the bases provides a blueprint (through the medium of another molecule, RNA) for the proteins that it constructs. Three letters, each representing a base, make up the code for a particular amino acid that, when joined with a string of others, creates a protein. Santa, like all humans, has about 3,000 million bases in his genetic makeup, or genome, but only around 70,000 genes that work to make proteins.

ENTROPY: A quantity that determines a system's capacity to evolve irreversibly in time, describing, for instance, the propensity of snowflakes to melt. Loosely speaking, we may also think of entropy as measuring the degree of randomness or disorder in a system.

ENZYME: The word *enzyme,* which denotes the huge proteins that cells use to transform other molecules, was coined from the Greek words for "in yeast." An enzyme is a biological catalyst, usually comprising a large protein molecule that accelerates essential chemical reactions in living cells.

EVOLUTION: From the Latin *evolutio,* "unfolding," evolution is the idea that all creatures — humans, reindeer, and robins — share the same line of descent. Proposed in its modern form by Charles Darwin (1809–1882), who, in the book *On the Origin of Species* (1859), also suggested a mechanism: inherited diversity, a struggle for existence that means that not all those born can survive and pass on their heritage; and natural selection, inherited differences in the chances of reproduction. Variants that increase their carrier's ability to make copies of themselves hence become more common;

those that hinder it become rarer. In time, Darwin suggested, this mechanism leads to the evolution of new forms of life — the origin of species. Darwin's ideas fit perfectly with those of modern genetics. Diversity arises through mutation (random changes in DNA) from generation to generation.

FRACTAL GEOMETRY: From the Latin *fractus*, "broken." The geometry used to describe an irregular shape that appears the same on all scales, whether you look at it close up or from far away. Fractals display the characteristic of self-similarity, an unending series of motifs within motifs repeated at all length scales. They abound in nature — for example, in clouds and snowflakes.

FULLERENES: The 1985 discovery of these new forms of carbon, distinct from carbon and graphite, amazed chemists, because carbon compounds had been intensively studied for decades without a hint of the existence of this branch of chemistry. The most famous fullerene, found in flickering candle flames, is buckminsterfullerene, or the buckyball, a molecule consisting of sixty carbon atoms. This football-shaped molecule with twenty hexagonal surfaces and twelve pentagonal surfaces, reminded researchers of the geodesic domes designed by the American architect R. Buckminster Fuller, hence its name.

FUZZY LOGIC: In mathematics and computing, a form of knowledge representation suitable for imprecise notions — such as cold, loud, and Christmassy — that depend on their contexts.

GAME THEORY: A branch of mathematics that deals with strategic problems, such as those that arise in business, commerce, evolution, and warfare, by assuming that the organisms involved invariably try to win. It can be applied equally well to Christmas shopping.

GENE: A unit of heredity comprising the chemical DNA, which is responsible for passing on specific characteristics from parents to offspring, such as a propensity to pile on the pounds after overeating. Each gene is the blueprint for a protein.

GENETIC CODE: The sequence of chemical building blocks of DNA (bases) that spells out the instructions to make amino acids, the building blocks of proteins.

GENETIC ENGINEERING: Tinkering with the genetic code (*see* DNA) of a creature to produce animals and plants with desirable properties — for example, Christmas trees that don't shed their needles easily.

GLUCOSE: A sugar present in blood that is the source of energy for the body.

HO, HO, HO: Scientists don't understand why Santa laughs, but they can at least define one. It begins as a deep breath. There follows a series of spasmodic, involuntary outward expirations, controlled by the opening and shutting of the glottis, the pathway from the throat to the lungs. The muscles then act on the larynx, lengthening or shortening the vocal cords to create different tones out of the explosions of air. The head and neck together act as an instrument to modify the sound.

IMMUNE SYSTEM: The range of weapons at the body's disposal to fight foreign forces, such as bacteria, viruses, and fungi.

MAGNETIC RESONANCE IMAGING (MRI): A non-invasive and painless way to examine and depict the interior of the body, even its metabolism. It can also be used to find a sixpence in a Christmas pudding.

MAILLARD REACTION: A chemical reaction between carbohydrates and the amino acids of proteins that is responsible for the browning of certain cooked foods, such as turkeys.

MEGA: A prefix denoting multiplication by a million.

MOLECULAR BIOLOGY: The study of the molecular basis of life, including the biochemistry of molecules such as DNA and RNA.

MUTATION: A change in genes produced by a chance or deliberate change in the DNA that makes up the hereditary material of an organism. For example, a mutation in a gene may be one reason Santa is so fat.

NANOTECHNOLOGY: The building of devices on a molecular scale. (*Nanos* is Greek for "dwarf.") Some people believe that Santa exploits this technology to make presents.

NEURAL NETWORKS: Computers that can learn. They are loosely modeled on the vast interconnected networks of nerve cells (neurons) in the brain.

NEURON: The nerve cell that is the fundamental signaling unit of the nervous system.

NEUROTRANSMITTER: A chemical that diffuses across a synapse and thus transmits signals between nerve cells. A slowdown in neurotransmitter release may be the reason for slurred speech, clumsiness, slow reflexes, and loss of inhibitions in seasonal revelers who have consumed too much alcohol.

Nucleic acid: A complex organic acid made up of a long chain of units called nucleotides. The two types, known as DNA and RNA, form the basis of heredity.

Occultation: The passage of one astronomical object in front of another, obscuring it, as, for example, when the moon moves between Earth and Jupiter, hiding Jupiter from view. This kind of event may have been the sign the Wise Men were looking for to signify the birth of Jesus.

Phase change: A change in the physical state of material — for example, from solid snowmen to liquid water, liquid to vapor, or ice to vapor.

Pheromone: A chemical substance secreted by an animal that influences the behavior of other animals.

Photosynthesis: Probably the most important chemical reaction on Earth, because it enables Christmas trees and other plants to trap energy from the sun and convert it into a form that can sustain living cells. The key product of photosynthesis, which takes place in structures called chloroplasts, is the chemical adenosine triphosphate, the fuel for all the cell's activities. Eventually that energy fuels creatures higher up the food chain, such as human beings.

Polymers: Enormously versatile molecules found throughout nature, as well as in synthetic compounds with numerous commercial applications. They consist of long chains of atoms, and their physical versatility is derived from the variety in type, number, and arrangement of those atoms. The chain is constructed from small repeating units, or monomers.

Lignin (which makes Christmas trees woody), starch in potatoes, and the cellulose in paper decorations are all examples. Perhaps the most important polymer is DNA, the genetic blueprint for living organisms.

PROTEIN: A class of large molecules found in living organisms, consisting of strings of amino acids folded into complex but well-defined three-dimensional structures. For example, the proteins myosin and actin make up the fibers that give turkey muscle its texture.

QUANTUM GRAVITY: A much sought-after theory of everything that would straddle the world of the very small, as described by quantum theory, and the very big, as described by Einstein's theory of gravity.

QUANTUM THEORY: The most revolutionary scientific theory of the century, one that has to be taken into account in lasers, microelectronics, and Advent candles. It was pioneered between 1900 and 1928, mainly by European physicists who realized that the previous classical theories did not work when applied to subatomic particles. The reason is that only at this microscopic level does it matter that energy changes in abrupt and tiny jumps (quantum leaps). You can see when electrons make these jumps during fireworks displays or in candle flames. Red corresponds to a small quantum leap and blue to a relatively big one.

REACTION: In chemistry, the coming together of atoms or molecules, with the result that a chemical change takes place to rearrange atoms.

RECEPTORS: Triggering devices in the membranes of our cells that allow drugs to work. Hormones and drugs are like the keys that fit the receptors' locks. When a chemical activates a given receptor, it will trigger a specific response. For example, sex hormones activate their receptors to determine whether a person is a boy or a girl. A locally acting hormone called a prostaglandin sends the message to deposit fat by docking with a receptor when we eat too much at Christmas.

REDUCTIONISM: A doctrine according to which complex phenomena can be explained in terms of something simpler. In particular, atomistic reductionism contends that macroscopic phenomena, such as the heady aroma of mulled wine, can be explained in terms of the properties of atoms and molecules.

RELATIVITY: This theory deals with the concepts of space, time, and matter and was developed by Albert Einstein (1879–1955) as an extension of the ideas first set out by Sir Isaac Newton. The effects relativity predicts lie at the boundaries of our experience, in the domains of the supersmall, the superfast, the superlarge, and the supermassive. *Special relativity,* announced in 1905, starts from the premise that the speed of light and the laws of physics are the same — that is, invariant — for observers moving at constant speeds relative to one another.

General relativity, unveiled a decade later, broadens this idea so that the laws of physics should be the same for all observers, regardless of how they are moving relative to one another. Conceived after Einstein realized that a man falling off a roof would not feel his own weight, general relativity describes gravity not as a force, as Newton had suggested, but as the curvature of space-time, a four-dimensional mix of space and time. Relativity is now more relevant than ever, given the interest in explaining how Santa delivers all those presents on Christmas Eve.

RNA (RIBONUCLEIC ACID): The genetic material used to translate DNA into proteins. In some viruses, it can also be the principal genetic material.

SECOND LAW OF THERMODYNAMICS: The writer C. P. Snow said that it should be part of the intellectual complement of any well-educated person. Put baldly, the law states that a quantity called entropy always increases in an isolated system. This is a fancy way of saying that water molecules in a puddle do not spontaneously arrange into a snowman.

SNOW: Precipitation in the form of small ice crystals, which may fall singly or in tangled masses called flakes. The crystals are formed in clouds from water vapor.

SOLSTICE: One of the two times of the year when the sun reaches its greatest excursion north (summer solstice) and south (winter solstice) of the equator.

STATISTICAL MECHANICS: The discipline that attempts to express the properties of macroscopic systems in terms of their atomic and molecular constituents.

SUPERCOMPUTER: The fastest, most powerful type of computer.

TERA: A prefix indicating one million million.

THERMODYNAMICS: The science of heat and work.

UPC: The most ubiquitous sequence of black-and-white stripes on Earth. The UPC is the Universal Product Code, better known as the bar code. The pattern of bars and spaces can be read by a scanning device connected to a computer and is now widely used to keep track of stock, from food to books to records, and to help monitor the surge in sales that takes place every Christmas.

VIRTUAL REALITY: An advanced form of computer simulation in which a participant is plunged into an artificial environment.

VIRUS: One of the smallest infectious agents, consisting of a piece of genetic code wrapped in protein, measuring between fifteen and three hundred nanometers across (one nanometer is one billionth of a meter). Viruses are responsible for a huge range of diseases, such as influenza and that cold that often seems to strike around Christmas. It is debatable whether viruses are living, since they have to hijack the molecular machinery of our cells to reproduce. (They do this by "reprogramming" our cells with their genetic code, turning them into virus factories.) That is why the common cold is so tricky to treat: it is difficult to combat viruses with drugs without also harming the cells they parasitize, although safe, effective drugs do exist for some viral infections.

Y CHROMOSOME: The essence of human maleness. Inheritance of this chromosome, one of the bundles of DNA found in our cells, is what determines that a human embryo will become a boy. It is paired with the other sex chromosome, the X, which carries far more information. Men are XY and women are XX.

BIBLIOGRAPHY

Addinall, Peter. "A Response to R. J. Berry on 'The Virgin Birth of Christ.'" *Science & Christian Belief* 9, no. 1 (1997): 65–72.

Atkins, Peter. *Molecules.* New York: Scientific American Library, 1987.

Belk, Russell. "It's the Thought That Counts: A Signed Diagraph Analysis of Gift-Giving." *Journal of Consumer Research* 3 (1976): 155–162.

———. "Gift-Giving Behavior." *Research in Marketing* 2 (1979): 95–126.

Bentley, W. A., and W. J. Humphreys. *Snow Crystals.* 1931. Reprint, New York: Dover, 1962.

Bentley, W. A., and G. H. Perkins. "A Study of Snow Crystals." *Appleton's Popular Science,* May 1898, 75–82.

Berry, Roger. "A Response to P. Addinall." *Science & Christian Belief* 9, no. 1 (1997): 73–78.

———. "The Virgin Birth of Christ." *Science & Christian Belief* 8, no. 2 (1996): 101–110.

Björntorp, Per. "Obesity." *Lancet* 350, (1997): 423–426.

Blanchard, Duncan. "Wilson Bentley, Pioneer in Snowflake Photomicrography." *Photographic Applications in Science, Technology and Medicine* 8, no. 3 (May 1973): 26–28, 39–41.

Bonython, Elizabeth. *King Cole: A Picture Portrait of Sir Henry Cole, KCB, 1808–1882.* London: Victoria and Albert Museum.

Boyer, Bryce. "Christmas Neurosis." *Journal of the American Psychoanalytic Association* 3, no. 3 (1955): 467–488.

Braun, J., A. Glebov, A. P. Graham, A. Menzel, J. P. Toennies, "Structure and Phonons of the Ice Surface," *Physical Review Letters* 80 (1998): 2638.

Bulmer-Thomas, Ivor. "The Star of Bethlehem." *Quarterly Journal of the Royal Astronomical Society* 3, no. 4 (1992): 363–374.

Butts, Robert. *William Whewell's Theory of Scientific Method.* Pittsburgh: University of Pittsburgh Press, 1968.

Caplow, T. "Christmas Gifts and Kin Network." *American Sociological Review* 47 (1982): 383–392.

Carslaw, H. S., and J. C. Jaeger. *Conduction of Heat in Solids.* 2d ed. Oxford: Oxford University Press, 1959.

Cheal, D. *The Gift Economy.* New York: Chapman & Hall, 1988.

Chown, Marcus. "O Invisible Star of Bethlehem." *New Scientist* 148, no. 2009/2010 (1995): 34–35.

Clayton, Chris. "Bethlehem's Star." *Astronomy Now,* December 1996, 57–59.

Day, Peter, and Richard Catlow. *The Candle Revisited.* Oxford: Oxford University Press, 1994.

de Courcy, Geraldine. *Christmastide in Germany.* Bonn: Inter Nationes, 1957.

Dickens, Charles. *The Christmas Books.* Vol. 1. London: Penguin Books, 1985.

Dolara, Piero, Cristina Luceri, Caria Ghelardini, Claudia Monserrat, Silvia Aiolli, Francesca Luceri, Maura Lodovici, Stefano Menichetti, and Maria Novella Romanelli. "Analgesic Effects of Myrrh." *Nature* 379, no. 6560 (1996): 29.

East, Robert. *Consumer Behaviour: Advances and Applications in Marketing.* Hemel Hempstead, England: Prentice-Hall, 1997.

Emsley, John. *The Consumer's Good Chemical Guide.* Oxford: W. H. Freeman, 1994.

———. *Molecules at an Exhibition.* Oxford: Oxford University Press, 1998.

Epstein, David, and Raphael. "Bentley's Magnificent Obsession." *National Wildlife*, December 1963–January 1964, 32–34.

Faraday, M. *The Chemical History of a Candle*. London: Chatto & Windus, 1908.

Ford, Brian. "Even Plants Excrete." *Nature* 323, no. 6090 (1986): 763.

Furnham, Adrian. "Beware of Relations Bearing Gifts: They May Be Trying to Tell You Something You'd Rather Not Know." *New Scientist* 120, no. 1644 (1988): 80.

Furst, P. *The Encyclopaedia of Psychoactive Drugs: Mushrooms: Psychedelic Fungi*. London: Burke Publishing, 1986.

Golby, J. M., and A. W. Purdue. *The Making of Modern Christmas*. London: B. T. Batsford, 1986.

Gotoda, Takanari. "Born in Summer?" *Nature* 377, no. 6551 (1995): 672.

Greenberg, Leon. "Alcohol in the Body." *Scientific American* 189, no. 6 (1953): 86–90.

Hagstrom, Warren. "What Is the Meaning of Santa Claus?" *American Sociologist* 1 (1966): 248–254.

Halvorsen, Odd. "Epidemiology of Reindeer Parasites." *Parasitology Today* 2, no. 12 (1986): 334–339.

Hapgood, Fred. "When Ice Crystals Fall from the Sky Art Meets Science." *Smithsonian* 6, no. 10 (1976): 67–73.

Harrison, Albert, Nancy Struthers, and Michael Moore. "On the Conjunction of National Holidays and Reported Birthdates: One More Path to Reflected Glory?" *Social Psychology Quarterly* 51, no. 4 (1988): 365–370.

Hillier, Bevis. *Greetings from Christmas Past*. London: Herbert Press, 1982.

Holmes, Michael. "Revolutionary Birthdays." *Nature* 373, no. 6514 (1995): 468.

Hughes, David. *The Star of Bethlehem Mystery*. London: Dent, 1979.

Humphreys, Colin. "The Star of Bethlehem." *Science & Christian Belief* 5, no. 2 (1993): 83–101.

Humphreys, Colin, and W. G. Waddington. "Dating the Crucifixion." *Nature* 306, no. 5945 (1983): 743–746.

Keathley, D. E. "Biological Enhancement of Christmas Tree Production in Michigan." *Michigan Christmas Tree Journal* 36 (1993): 38–40.

Keverne, Eric, Fran Martel, and Claire Nevison. "Primate Brain Evolution: Genetic and Functional Considerations." *Proceedings of the Royal Society London B* 262, (1996): 689.

Koenig, Harold. *Is Religion Good for Your Health? Effects of Religion on Physical and Mental Health.* New York: Haworth Press, 1997.

Koenig H. G., H. J. Cohen, D. G. Blazer, et al. "Religious Coping and Depression in Elderly Hospitalized Medically Ill Men." *American Journal of Psychiatry* 149 (1992): 1693–1700.

Koenig, H. G., H. J. Cohen, L. K. George, J. C. Hays, D. G. Blazer. "Attendance at Religious Services, Interleukin-6, and Other Biological Indicators of Immune Function in Older Adults." *International Journal of Psychiatry in Medicine* 27 (1997): 233–250.

Koenig, H. G., S. Ford, L. K. George, D. G. Blazer, and K. G. Meador. "Religion and Anxiety Disorder: An Examination and Comparison of Associations in Young, Middle-Aged, and Elderly Adults." *Journal of Anxiety Disorders* 7 (1993): 321–342.

Koenig, H. G., L. K. George, H. J. Cohen, J. C. Hays, D. G. Blazer, D. B. Larson. "The Relationship Between Religious Activities and Blood Pressure in Older Adults." *International Journal of Psychiatry in Medicine* (1998): in press.

Laroche, Michel, Chankon Kim, Gad Saad, and Elizabeth Browne. "Determinants of In-Store Information Search Strategies Pertaining to a Christmas Gift Purchase." Working paper, Concordia University, 1997.

Leader-Williams, N. *Reindeer on South Georgia.* Cambridge: Cambridge University Press, 1988.

Martin, W. T. "Religiosity and United States Suicide Rates, 1972–1978." *Journal of Clinical Psychology* 40, no. 5 (1984): 1166–1169.

Matthews, Robert. "Odd Socks: A Combinatoric Example of Murphy's Law." *Mathematics Today,* March–April 1996, 39–41.

———. "Hurry Up and Wait." *New Scientist,* July 19, 1997, 24–27.

McCullough, M. E., D. B. Larson, H. G. Koenig, M. G. Milano. "Systematic Review of Published Research on Religious Commitment and Mortality." *Journal of the American Medical Association* (1997): in submission.

McElduff, Patrick, and Annette Dobson. "How Much Alcohol and How Often? Population Based Case-Controlled Study of Alcohol Consumption and Risk of a Major Coronary Event." *British Medical Journal* 314 (1997): 1159–1164.

McGee, Harold. *On Food and Cooking: The Science and Lore of the Kitchen.* New York: Collier Books, 1984.

Miller, Daniel. *Unwrapping Christmas.* Oxford: Clarendon Press, 1995.

Molnar, M. R. "An Explanation of the Christmas Star Determined from Roman Coins of Antioch." *Celator* 5, no. 12 (1991): 8–12.

———. "The Coins of Antioch." *Sky & Telescope* 83 (1992): 37–39.

———. "The Magi's Star from the Perspective of Ancient Astrological Practices." *Quarterly Journal of the Royal Astronomical Society* 36 (1995): 109–126.

Montague, Carl, Sadaf Farooqi, Jonathan P. Whitehead, Maria A. Soos, et al. "Congenital Leptin Deficiency Is Associated with Severe Early-Onset Obesity in Humans." *Nature* 387, no. 6636 (1997): 903–908.

Moore, Peter. "Why Be an Evergreen?" *Nature* 173, no. 5996 (1984): 703.

Morris, Desmond. *Christmas Watching.* London: Jonathan Cape, 1992.

Mullet, Mary B. "The Snowflake Man." *American Magazine* 99 (1925): 28–31.

Nittmann, J., and H. E. Stanley. "Connection Between Tip-splitting Phenomena and Dendritic Growth." *Nature* 321 (1986): 663–668.

Nittmann, J., and H. E. Stanley. "Non-Deterministic Approach to Anisotropic Growth Patterns with Continuously Tunable Morphology: The Fractal Properties of Some Real Snowflakes." *Journal of Physics* 20 (1987): L1185.

North, Adrian, and David Hargreaves. "The Musical Milieu: Studies of Listening in Everyday Life." *Psychologist* 10, no. 7 (1997): 309–312.

Ohlsson, R., K. Hall, and M. Ritzen. *Genomic Imprinting: Causes and Consequences.* Cambridge: Cambridge University Press, 1995.

Otnes, C., Y. Kim, and T. Lowrey. "Ho, Ho, Woe: Christmas Shopping for 'Difficult' People." In *Advances in Consumer Research,* vol. 19, ed. Sherry and Sternthal. Provo, Utah: Association for Consumer Research, 1992.

Oxman, T. E., D. H. Freeman, and E. D. Manheimer. "Lack of Social Participation or Religious Strength and Comfort as Risk Factors for Death After Cardiac Surgery in the Elderly." *Psychosomatic Medicine* 57 (1995): 5–15.

Pargament, Kenneth. *The Psychology of Religious Coping: Theory, Research, and Practice.* New York: Guilford Press, 1997.

Parkinson, Claire. *Breakthroughs: A Chronology of Great Achievements in Science and Mathematics, 1200–1930.* London: Mansell Publishing, 1985.

Pollay, R. "It's the Thought That Counts: A Case Study in Xmas Excesses." In *Advances in Consumer Research,* vol. 14, ed. Wallendorf and Anderson. Provo, Utah: Association for Consumer Research, 1986.

Pond, C. M. *The Fats of Life.* Cambridge: Cambridge University Press, 1988.

———. "An Evolutionary and Functional View of Mammalian Adipose Tissue." *Proceedings of the Nutrition Society* 51 (1992): 367–377.

———. "The Structure and Organization of Adipose Tissue in Naturally Obese Non-hibernating Mammals." In *Obesity in Europe '93: Proceedings of the Fifth European Congress on Obesity,* ed. H. Ditschuneit, F. A. Gries, H. Hauner, V. Schusdziarra, and J. G. Wechsler. London: J. Libbey, 1994.

Pond, C. M., C. A. Mattacks, R. H. Colby, and N. J. Tyler. "The Anatomy, Chemical Composition and Maximum Glycolytic Capacity of Adipose Tissue in Wild Svalbard Reindeer *(Rangifer tarandus platyrhynchus)* in Winter." *Journal of Zoology, London* 229 (1993): 17–40.

Pretl, George, Winnifred Cutler, Carol Christensen, Henry Lawley, George Higgins, and Celso-Raman Garcia. "Human Axillary Extracts." *Journal of Chemical Ecology* 13, no. 4 (1987): 717–731.

Proebsting, Bill, and Jose Montano. "Needle Abscission in Douglas Fir." *Christmas Tree Lookout* 23, no. 2 (1990): 30–32.

Rudgley, R. *The Alchemy of Culture.* London: British Museum Press, 1993.

Samuel, Delwen. "Archaeology of Ancient Egyptian Beer." *Journal of the American Society of Brewing Chemists* 54, no. 1 (1996): 3–12.

———. "Investigation of Ancient Egyptian Baking and Brewing Methods by Correlative Microscopy." *Science* 273 (1996): 488–490.

Sansom, William. *Christmas.* London: Weidenfeld & Nicolson, 1968.

Schatzman, Morton. "Does Christmas Drive You Crackers?" *New Scientist* 120, no. 1644 (1988): 46–48.

Sen, S., J. Aimers-Halliday, C. R. McKinley, and R. J. Newton. "Micropropagation of Conifers by Organogenesis." *Plant Physiology* 12 (1993): 129–135.

Sen, S., M. E. Magallanes-Cedeno, R. H. Kamps, C. R. McKinley, and R. J. Newton. "In Vitro Micropropagation of Afghan Pine." *Canadian Journal of Forestry Research* 24 (1994): 1248–1252.

Sherry, J., and M. McGrath. "Unpacking the Holiday Presence: A Comparative Ethnography of Two Gift Stores." In *Interpretive Consumer Research,* ed. E. Hirschmann. Provo, Utah: Association for Consumer Research, 1989.

Simons, Paul. *Weird Weather.* London: Little, Brown, 1996.

Singer, Charles, E. J. Holmyard, A. R. Hall, and T. I. Williams. *A History of Technology.* Vol. 4. Oxford: Clarendon Press, 1958.

Smith, Sheryl, Qui Hua Gong, Fu-Chan Hsu, Ronald Markowitz, J. Ffrench Mullen, Xin She Li. "GABA Receptor Alpha-4 Subunit Suppression Prevents Withdrawal Properties of an Endogenous Steroid." *Nature* 392, no. 6679 (1998): 926–930.

Smith, T. K., S. R. R. Musk, and I. T. Johnson. "Allyl Isothiocyanate Selectively Kills Undifferentiated HT29 Cells in Vitro and Suppresses Aberrant Crypt Foci in the Colonic Mucosa of Rats." *Biochemical Society Transactions* 24 (1996): 381.

Sterba, Richard. "On Christmas." *Psychoanalytic Quarterly* 13 (1944): 79–83.

Stern, Kathleen, and Martha McClintock. "Regulation of Ovulation by Human Pheromones." *Nature* 392, no. 6672 (1998): 177.

Stoddard, Gloria May. *Snowflake Bentley: Man of Science, Man of God.* St. Louis: Concordia Publishing, 1979.

Verhagen, Hans, H. E. Poulsen, S. Lott, G. van Poppel, et al. "Reduction of Oxidative DNA Damage in Humans by Brussels Sprouts." *Carcinogenesis* 16, no. 4 (1995): 969–970.

Vines, Gail. "My Best Friend's a Brussels Sprout." *New Scientist* 152, no. 2061 (1996): 46–49.

Yeo, Richard. *Defining Science: William Whewell: Natural Knowledge, and Public Debate in Victorian Britain.* Cambridge: Cambridge University Press, 1993.

INDEX